AF541552

Technology Commercialization and Incubation

NIPA® GENX ELECTRONIC RESOURCES & SOLUTIONS P. LTD.
New Delhi-110 034

About the Authors

Professor Kausik Pradhan, is an erudite personality in the academic world of Agricultural Extension with his constructive thought, extreme mettle of innovative ideas, research acumen and robust diligence in academics. Presently he is working as a Professor, the Department of Agricultural Extension, Uttar Banga Krishi Viswavidyalaya, West Bengal, India. He has worked as a key trainer in different training programmes on participatory extension and group dynamics as well as strengthening Farmers' Producer Organisations. He has associated with more than six (6) mega nationally and internationally funded research and development projects including the ICAR-National Agricultural Science Fund project. He has developed his expertise in the areas of Technology Socialisation, Women empowerment, Micro planning, Entrepreneurship management, Natural Resource Management, Value addition and Supply Chain Management, Participatory Extension, Market-led Extension, need-based technology delivery model through Farmers' Producer Organisations. The need-based extension delivery models through Farmers' Producer Organisations are acknowledged by Indian Council of Agricultural Research as Innovative technologies. He has also been an integral part of several trainings, workshops and conferences on Food System Innovation, Sustainable Resilient Farming Systems in Australia, Nepal and Bangladesh. Presently he is serving as the Chief Editor of the prestigious journal, Indian Research Journal of Extension Education. He served as the Associate Editor of more than five National and International Journals. To his credit he has published so far 147 (one hundred forty-seven) numbers of research papers in reputed national and international journals, contributed 27 (twenty-seven) chapters in different renowned books and co-authored 10 (ten) books. Different national-level academic societies also recognize his contribution by conferring awards namely Young Scientist Award, Science Excellence Award, Best Extension Professional Award and many more. He is the prestigious Fellow of International Society for Research and Development, Scholars Academic and Scientific Society and Society of Extension Education.

Chigilipalli Mounika holds a B.Sc. (Hons.) in Horticulture from Dr. YSR Horticultural University, V.R. Gudem, and an M.Sc. in Agricultural Extension from Vasantrao Naik Marathwada Krishi Vidyapeeth, Parbhani. She qualified the ASRB NET in Agricultural Extension in 2023. Currently, she is pursuing a Ph.D. in Agricultural Extension at Uttar Banga Krishi Viswavidyalaya (UBKV), Pundibari, West Bengal. She has actively participated in numerous national and international (online and offline) seminars, conferences, and workshops, where she has presented her research findings. Her academic contributions include the publication of book chapters, popular articles, meeting note and research papers in reputed journals, particularly in the domains of agricultural Extension, soft skills and climate change. She has also undergone various trainings focused on statistical tools, software applications, and recent advancements in Agricultural Extension.

Seepana Anil Kumar is a Ph.D. Research Scholar in the Department of Agricultural Extension Education, College of Agriculture, Rajendranagar, Professor Jayashankar Telangana State Agricultural University, Telangana. He completed his B.Sc. (Hons.) in Horticulture from Dr. Y.S.R. Horticultural University, Venkataramannagudem, West Godavari District, Andhra Pradesh in 2020. He pursued his M.Sc. in Agricultural Extension Education from Vasantrao Naik Marathwada Krishi Vidyapeeth, Parbhani, Maharashtra, and graduated in 2022. He has qualified ASRB-NET (2023) and UGC-NET (June and December 2024) in Adult Education. He has contributed to the academic community through the publication of research papers, book chapters, and popular articles in reputed national and international journals.

Dr. Indrajit Barman is working as Assistant Professor at Assam Agricultural University. He served ICSSR as well as ICAR Institutes for more than 9 years. He qualifies ASRB NET for three times. He was also selected for PhD Fellowship from ICSSR, New Delhi. He Served 6 years In Krishi Vigyan Kendra as Subject Matter Specialist (Agril. Extension). Moreover, He also worked with The Energy and Research Institute (TERI) for one year. He obtained his Ph. D. degree qualifying ASRB SRF examination from Uttar Banga Krishi Viswavidyalaya, Cooch Behar, West Bengal. His publication covers a number of Research articles, Edited Books, Book Chapters, Training Manuals, Instructional Manuals and extension literature such as leaflet, folder, booklet, pamphlet etc. He teaches both UG and PG students of university and guides a number PG student for their PG research work. The Co-Author also participated in many International and National seminar/Conferences/Workshops and presented his research findings. He is also member of many professional organizations/Societies.

Shashikant, Working as Assistant Professor Cum Junior scientist, at Nalanda college of Horticulture Noorsari, Nalanda in Department of Extension Education under Bihar Agricultural University, Sabour Bhagalpur. He had 12 years of experience of Teaching, Research, Extension. He had conducted more than five research projects and wrote 5 books as author and 35 research publications in NAAS rated journals and more than 100 articles in different publications. He has more experience in ICT and give training of crop improvement via ICT tools.

Technology Commercialization and Incubation

Kausik Pradhan
Chigilipalli Mounika
Seepana Anil Kumar
Indrajit Barman
Shahsikant

NIPA® GENX ELECTRONIC RESOURCES & SOLUTIONS P. LTD.
New Delhi-110 034

NIPA® GENX ELECTRONIC RESOURCES & SOLUTIONS P. LTD.

101,103, Vikas Surya Plaza, CU Block
L.S.C.Market, Pitam Pura, New Delhi-110 034
Ph : +91 11 4386 0225, 9717133558, 9540816132
E-mail: newindiapublishingagency@gmail.com
Website: www.niparesources.com

Print ISBN: 978-93-72193-72-5

ebook ISBN: 978-93-72199-49-9

Composed and Designed by NIPA®.

উত্তরবঙ্গ কৃষি বিশ্ববিদ্যালয়
পুণ্ডিবাড়ী, কোচবিহার , পশ্চিমবঙ্গ-৭৩৬১৬৫
UTTAR BANGA KRISHI VISWAVIDYALAYA
P.O. PUNDIBARI, DIST. COOCH BEHAR, WEST BENGAL- 736165

প্রফেসর দেবব্রত বসু
উপাচার্য
Prof. Debabrata Basu
M.Sc. (Ag.), Ph.D., DDE
Vice-Chancellor

মো॰/Mob: +91-9434748016
ইমেল/E-mail: vc@ubkv.ac.in
vcubkvv@gmail.com
ওয়েবসাইট/ Website: www.ubkv.ac.in

Ref. No. VC/UBKV- 4120

Date: 23 June 2025

Foreword

Innovation and entrepreneurship are driving forces behind today's global progress. In this context, the commercialization of technology and the support provided through incubation have become vital tools for economic growth and social development. This book, Technology Commercialization and Incubation, is both timely and relevant, as it highlights how research outcomes can be turned into practical solutions and how start-ups can be nurtured into successful ventures.

Technology commercialization is not just about moving ideas from the lab to the market. It is a complex process that involves innovation, market understanding, policy support, and teamwork among various stakeholders. Key aspects such as protecting intellectual property, understanding customer needs, navigating regulations, and developing viable products are all part of this journey. When done right, commercialization can transform innovative ideas into meaningful impacts.

Incubation plays a crucial role by providing a supportive environment for start-ups in their early and often difficult stages. Incubators offer infrastructure, mentorship, funding, technical support, and networking opportunities. They help entrepreneurs take risks, learn from failures, and turn their ideas into sustainable businesses. Incubation is especially important for youth, rural innovators, and first- generation entrepreneurs.

This book is a valuable resource for students, researchers, faculty members, policymakers, and entrepreneurs. It combines insights from multiple disciplines including science, management, law. and public policy. It also highlights the important role of academic institutions and research organizations in promoting innovation through incubation and technology transfer.

Aligned with the new BSMA syllabus, this book is especially useful for Ph.D. scholars in Agricultural Extension and related fields. It covers a wide range of topics—from the basics of innovation and intellectual property to grassroots innovations, policy frameworks, and start-up ecosystems. It also discusses key government initiatives such as Start-up India, Make in India, Digital India, and the Atal Innovation Mission.

As the world faces major challenges like climate change, food insecurity, and digital inequality, promoting innovation and entrepreneurship has never been more important. I believe this book will inspire and guide its readers to contribute meaningfully in building a more inclusive and innovation- driven future.

Prof. Debabrata Basu
Vice-Chancellor, UBKV

Preface

In the rapidly evolving landscape of innovation and entrepreneurship, the journey from a laboratory discovery to a viable market product is both exciting and complex. This book, Technology Commercialization and Incubation, is designed to bridge the critical gap between academic research and its successful application in the commercial world. With the ever- growing emphasis on translating scientific breakthroughs into socially and economically beneficial technologies, this book serves as a timely resource for scholars, innovators, policymakers, and institutions.

This book is especially curated to meet the academic and professional needs of Ph.D. students, particularly those involved in agricultural, engineering, life sciences, and management disciplines. It provides a comprehensive framework that aligns with the new guidelines set by the Board of Studies in Multidisciplinary Approaches (BSMA) under the Indian Council of Agricultural Research (ICAR). The BSMA's revised curricula emphasize not only academic excellence but also skill-based, experiential learning that fosters innovation, incubation, entrepreneurship, and technology transfer. In line with these reforms, this book aims to build capacity in understanding and navigating the process of commercialization, intellectual property management, startup ecosystems, and institutional incubation mechanisms.

The chapters of this book cover a wide array of topics relevant to modern technology incubation and commercialization. These include fundamentals of innovation, types and stages of incubation, role of technology business incubators (TBIs), intellectual property rights and valuation, legal and regulatory frameworks, funding mechanisms, startup acceleration programs, market assessment strategies, and public-private partnerships. Real-world case studies, recent policy initiatives, and success stories from premier Indian institutions and startups add practical depth to the theoretical foundations. One of the core strengths of this book is its interdisciplinary and application-oriented approach. Each chapter is written in a lucid style, integrating theoretical frameworks with field-based insights, making the content accessible and relevant. Furthermore, this book reflects the national priorities of Atmanirbhar Bharat (self-reliant India), Startup India, and Make in India missions. It encourages young researchers and Ph.D. scholars to not only focus on academic excellence but also to think entrepreneurially, contributing to innovation-led economic growth and societal development.

We hope this book will serve as a foundational text and a source of inspiration for young scholars, faculty members, innovation managers, and ecosystem enablers who aim to transform research into real-world impact. By fostering a deeper understanding of technology commercialization and incubation, this book aspires to contribute meaningfully to India's growing innovation and startup ecosystem.

Authors

Contents

Abbreviations

ABS: Access and Benefit Sharing

AI: Artificial Intelligence AIM: Atal Innovation Mission AO: Appellation of Origin

BIMSTEC: Bay of Bengal Initiative for Multi-Sectoral Technical and Economic Cooperation

CBD: Convention on Biological Diversity CBRs: Community Biodiversity Registers DCF: Discounted Cash Flow

DPIIT: Department for Promotion of Industry and Internal Trade DPIIT: Department for Promotion of Industry and Internal Trade DUS: Distinctiveness, Uniformity, and Stability

ECAST: Expert and Citizen Assessment of Science and Technology

GEAC: Genetic Engineering Approval Committee

GFIs: Grassroots and farmers' innovations

GI: Geographical Indication

GMOs: Genetically Modified Organisms

IoT: Internet of Things

IPO: Indian Patent Office

IPR: Intellectual Property Rights

ITK: Indigenous Traditional Knowledge

LBMCs: Local Biodiversity Management Committees

MAT: Mutually Agreed Terms

NAIP: National Agricultural Innovation Project

NEP: National Environment Policy

NGOs: Non-Governmental Organizations

NSTEDB: National Science & Technology Entrepreneurship Development Board

OECD: Organisation of Economic Co-operation and Development

PBRs: People's Biodiversity Registers

PIC: Prior Informed Consent

PPV&FR: Protection of Plant Varieties and Farmers' Rights

PRA: Participatory Rural Appraisal

PTA: Participatory technology Assessment RKVY: Rashtriya Krishi Vikas Yojana ROI: Returns on Investments

SRISTI: Society for Research and Initiatives for Sustainable Technologies and Institutions

TAI: Technology Assessment Index

TBI: Technology Business incubator

TKDL: Traditional Knowledge Digital Library

TRIPS: Trade-Related Aspects of Intellectual Property Rights

TTO: Technology Transfer Office

UNCBD: United Nations Convention on Biological Diversity

UNEP: United Nations Environment Programme

UNFCCC: United Nations Framework Convention on Climate Change UPOV: International Union for the Protection of New Varieties of Plants USTR: U.S. Trade Representative

WTO: World Trade Organization

1

Technology Commercialization and Modern Concept

Basics of Technology Commercialization

Technology, a ubiquitous force shaping our world, can be understood from various angles. Here's a breakdown of its definition, functions, and the fascinating process of its advancement.

Definition

At its core, technology is the application of scientific knowledge for practical purposes. It's not just about gadgets and gizmos; it encompasses a vast array of systems, processes, and tools that we use to transform the world around us and solve problems. This includes everything from the humble wheel to the intricate workings of artificial intelligence.

Functions

Technology serves a multitude of functions, broadly categorized into:

Simplifying tasks: From automated manufacturing to self-driving cars, technology automates and streamlines processes, saving time and effort.

Enhancing communication and connection: From the internet to social media, technology bridges geographical barriers and fosters global interaction.

Expanding our knowledge and understanding: Scientific instruments, telescopes, and advanced data analysis tools enable us to explore the universe, delve into the microscopic world, and gain deeper insights into various fields.

Improving health and well-being: Medical technology advancements like prosthetics, gene editing, and telemedicine are revolutionizing healthcare and improving lives.

Entertainment and leisure: From virtual reality experiences to on-demand streaming services, technology provides diverse avenues for recreation and enjoyment.

Process of Technological Advancement

Technological progress is driven by a dynamic interplay of four key elements:

Invention: The spark of creativity that leads to the conception of a new idea or device. This could be a groundbreaking scientific discovery or a clever solution to a practical problem.

Discovery: The process of uncovering existing knowledge or phenomena that were previously unknown. Scientific research and exploration often lead to discoveries that pave the way for new technologies.

Innovation: The practical application of inventions and discoveries to create new products, processes, or services. This involves turning theoretical concepts into tangible solutions with real- world benefits.

Diffusion: The spread and adoption of new technologies by individuals, communities, and societies. This involves overcoming challenges like infrastructure limitations, social acceptance, and economic feasibility.

Understanding these elements and their interplay is crucial for appreciating the dynamic nature of technological advancement. It's a continuous cycle of curiosity, creativity, and ingenuity that propels us forward, shaping our present and defining our future.

The Four Pillars of Innovation

Innovation, the lifeblood of progress, manifests in various forms. Here's a breakdown of four key types, each with its unique approach to pushing boundaries and driving change:

1. Basic Research

Definition: Exploration of fundamental scientific concepts and principles without immediate practical application in mind. It's the quest for knowledge and understanding without preordained goals.

Impact: Lays the groundwork for future breakthroughs, providing the building blocks for new technologies and discoveries. It can spark unexpected applications years or decades later.

Examples: The discovery of electricity by Benjamin Franklin, the development of the theory of relativity by Albert Einstein.

2. Breakthrough Innovation

Definition: Creating entirely new products, services, or processes that revolutionize existing markets or create new ones. It's about challenging the status quo and offering radical solutions.

Impact: Disruptive and game-changing, it can reshape industries, create new jobs, and profoundly impact society.

Examples: The invention of the internet, the development of the first smartphone, the invention of gene editing technology.

3. Disruptive Innovation

Definition: Introducing new, simpler, and often cheaper solutions that initially target niche markets but eventually disrupt and displace established players in existing markets.

Impact: Creates competition, drives down costs, and offers more accessible solutions for consumers. Can lead to the downfall of established companies if they fail to adapt.

Examples: The rise of low-cost airlines, the emergence of e-commerce giants like Amazon, the adoption of mobile banking.

4. Sustaining Innovation

Definition: Incremental improvements to existing products, services, or processes to maintain a competitive edge and meet evolving customer needs. It's about continuous refinement and optimization.

Impact: Ensures long-term success by enhancing existing offerings, offering better performance, and satisfying changing consumer preferences.

Examples: The ongoing improvements in smartphone cameras, the development of new car models with increased fuel efficiency, the constant updates and features added to software applications.

5. Incremental Innovation

Incremental innovation involves the steady and ongoing enhancement of current products, services, or processes. Although it may not be as attention-grabbing as other forms of innovation, it delivers clear and consistent value, especially to well- established businesses. By continuously refining their offerings and operations, companies can avoid stagnation and steadily increase their share in the market.

6. Radical Innovation

Radical innovation often stems from a major technological advancement that disrupts existing industries and leads to the formation of entirely new markets. It significantly alters how a company engages with its market. The success of such transformative innovation often depends on the organization's internal culture, behaviors, and capabilities, which must support and enable the development and commercialization of groundbreaking ideas.

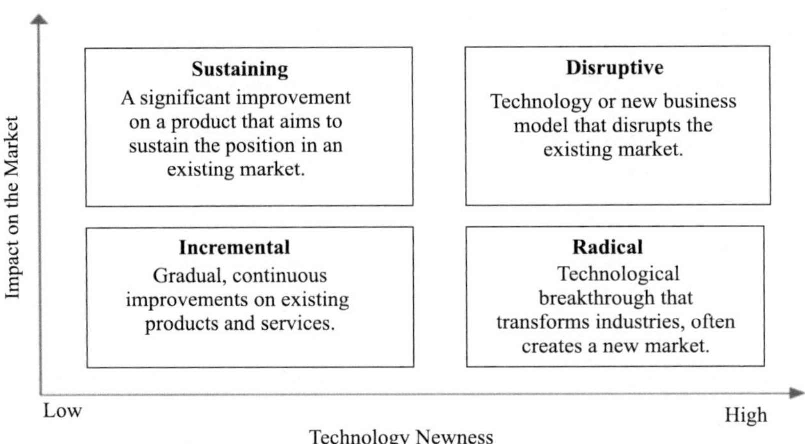

Understanding the interplay between these four types of innovation is crucial for businesses and individuals. While basic research provides the seeds of future breakthroughs, disruptive innovation can shake up established markets. Sustaining innovation ensures long-term success, while breakthrough innovation can redefine entire industries.

Remember, innovation is not a singular act but a continuous process driven by diverse approaches. Embracing these different types and fostering a culture of creativity are key to unlocking the potential of progress and shaping a better future.

Technology Transfer and Commercialization

Technology transfer and commercialization is the fascinating journey of taking scientific discoveries and innovations from the lab bench to the real world, turning them into products and services that benefit society. It's a process with immense potential to improve lives, create jobs, and drive economic growth. Technology transfer and commercialization represent the crucial bridge between the world of scientific discovery and the realm of practical applications. It's the process of transforming innovative ideas, often born in research labs and universities, into products and services that benefit society at large. Technology transfer and commercialization is a fascinating dance, bridging the gap between the world of scientific breakthroughs and the everyday products and services we use.

What is it?

Technology transfer: Moving technology from one entity (a researcher, university, lab) to another (a company, entrepreneur). It refers to the movement of technology from the entity that holds it (e.g., a research institution, inventor) to another entity capable of bringing it to market (e.g., a company, entrepreneur). This may include several different methods like licensing, patents, joint ventures, and startups.

Commercialization: Involves the development and marketing of a product or service based on the transferred technology. This requires significant effort in Activities involving market intelligence and product innovation, manufacturing, and sales. Successful commercialization ensures the technology reaches its intended audience and delivers its intended benefits.

Why is it Important?

Technology transfer and commercialization play a crucial role in:

Boosting economic growth: By turning research into marketable products, it creates jobs, stimulates investment, and generates revenue.

Solving societal challenges: From tackling climate change to improving healthcare, commercialized technologies can address critical problems facing society.

Improving quality of life: New products and services can enhance our daily lives, from communication and entertainment to healthcare and transportation.

Turns breakthroughs into practical solutions, addressing real-world challenges. Stimulates economic growth by creating new businesses and jobs.

Improves healthcare, education, and other sectors with innovative technologies.

How Does it Work?

Identifying promising technologies: Universities and research institutions have dedicated offices to assess the commercial potential of their research.

Protecting intellectual property: Patents, copyrights, and trademarks are used to secure ownership and incentivize investment.

Finding the right partners: Matching technologies with companies that have the resources and expertise to bring them to market.

Negotiating licenses and agreements: Ensuring fair compensation for inventors and researchers while making the technology accessible for commercialization.

Challenges and Opportunities

- **Bridging the gap between academia and industry:** Different cultures and risk appetites can create communication hurdles.
- **Funding and investment:** Early-stage technologies often require significant resources to be brought to market.
- **Regulations and intellectual property issues:** Navigating complex legal landscapes can be challenging.
- **The potential for positive impact:** Despite the challenges, successful technology transfer and commercialization can bring immense benefits to society.
- **Intellectual property (IP) protection:** Ensuring inventors and institutions receive fair compensation for their contributions.
- **Market uncertainties:** Identifying the right market for the technology and navigating potential risks.
- **Financial resources:** Securing funding for research, development, and commercialization activities.
- **Communication and collaboration:** Building bridges between researchers, entrepreneurs, and investors.

Examples of Successful Cases

The development of life-saving drugs and medical devices.

The rise of clean power generation methods like solar and wind power.

The creation of innovative communication tools like smartphones and the internet.

Benefits of Successful Technology Transfer and Commercialization

Economic growth: Creates jobs, stimulates investment, and boosts regional economies.

Improved quality of life: Leads to new products and services that address societal challenges in healthcare, energy, and other sectors.

Scientific advancement: Generates funds for further research and development, accelerating the pace of innovation.

The future of technology transfer and commercialization is bright. With increasing emphasis on innovation and collaboration, we can expect to see even more exciting breakthroughs and advancements that touch every aspect of our lives.

Overall, technology transfer and commercialization play a critical role in transforming the world around us. By bridging the gap between scientific breakthroughs and practical applications, it holds the potential to drive progress, improve lives, and shape a better future.

Technology transfer and commercialization will continue to play a critical role in shaping our future. By fostering a collaborative and supportive ecosystem, we can bridge the gap between groundbreaking ideas and practical solutions, ultimately driving progress and improving lives around the world.

Nature of Agricultural Technology

Agricultural Technology

Agricultural technology, also known as agritech, encompasses the application of cutting-edge tools and techniques to develop the efficiency, productivity, and sustainability of agricultural practices. From robots tending crops to drones monitoring field health, agritech is transforming the way we grow our food. Agricultural technology, often shortened to AgTech, refers to the application of technology in the agricultural sector. It encompasses a wide range of tools, techniques, and innovations aimed at improving farm efficiency, productivity, and sustainability. From high-tech robots to simple but effective sensors, AgTech is transforming the way we grow food.

At its core, agritech aims to address the challenges faced by modern agriculture, such as:

Increasing food demand: As the global population grows, we need to produce more food on less land and with fewer resources.

Climate change: Rising temperatures, droughts, and floods threaten crop yields and food security.

Resource scarcity: Water, arable land, and nutrients are becoming increasingly scarce, making efficient use of these resources crucial.

Labor shortages: Attracting and retaining skilled workers in rural areas is becoming increasingly difficult.

Agritech Offers a Multitude of Solutions to these Challenges, Helping Farmers

Increase yield and productivity: Precision agriculture techniques like targeted fertilization and irrigation optimize resource use and boost crop yields.

Reduce waste and losses: Automated harvesting and storage systems minimize post-harvest losses and improve food quality.

Monitor and manage crops: Drones equipped with sensors can collect data on field health, enabling farmers to identify potential problems early and take corrective action.

Optimize resource use: Smart irrigation systems use sensors to deliver water only when and where it's needed, conserving this precious resource.

Improve sustainability: Technologies like vertical farming and precision agriculture can minimize environmental impact and promote sustainable practices.

Types of Agricultural Technology

The diverse world of agritech encompasses a wide range of technologies, broadly categorized into:

- **Precision agriculture:** Leveraging data and sensors to optimize resource use and make informed decisions about planting, irrigation, fertilization, and pest control.
- **Automation and robotics:** Utilizing robots and drones for tasks like planting, harvesting, weeding, and monitoring crops, reducing reliance on manual labor.
- **Biotechnology:** Applying genetic engineering and other biological techniques to design and produce new plant cultivars with improved yield, disease resistance, and stress tolerance.
- **Artificial intelligence (AI) and machine learning:** Analyzing vast amounts of data to predict weather patterns, detect diseases, and make informed recommendations for optimizing farming practices.
- **Remote sensing and satellite imagery:** Monitoring crops from space to track growth, health, and water stress, enabling farmers to make timely interventions.

- **Digital platforms and apps:** Providing farmers with access to real-time data, market information, and tools for managing their operations efficiently.
- **Internet of Things (IoT):** Sensors and other devices are being used to connect farms to the internet, allowing for real-time monitoring of conditions and remote control of equipment. This data can be used to optimize farm operations and make informed decisions.
- **Vertical farming:** This technique involves growing crops in controlled environments, such as warehouses or rooftops. This can be beneficial in areas with limited land or harsh climates. These are just a few examples of the diverse and rapidly evolving landscape of agritech. As technology continues to advance, we can expect even more innovative solutions to emerge, transforming the way we feed the world.

By embracing agritech, farmers can not only improve their profitability and sustainability but also contribute to a more secure and resilient food system for future generations.

Technology Generation Systems

Technology, like a phoenix rising from the ashes of innovation, undergoes a fascinating journey from conception to obsolescence. Understanding this journey, characterized by the technology generation system and the technology life cycle, is crucial for navigating the ever-evolving landscape of our world.

Technology Generation Systems

These systems are the engines that drive the creation of new technologies. They encompass various mechanisms, each with its own strengths and weaknesses:

Research & Development (R&D): The traditional approach, where dedicated teams within companies or universities conduct research and development activities. This system is often resource-intensive but can lead to breakthrough innovations.

Open Innovation: This collaborative approach involves tapping into the collective intelligence of diverse stakeholders, including universities, startups, and even individual inventors. It fosters faster development and diverse perspectives but can be challenging to manage.

Serendipity and Accidental Discovery: Sometimes, the most groundbreaking innovations arise from unexpected encounters or chance discoveries. While not easily replicable, these serendipitous breakthroughs can rewrite the rules of the game.

Crowdsourcing and User-Driven Innovation: Leveraging the collective wisdom of the crowd through online platforms can generate innovative solutions based on real-world needs and feedback. This approach is often cost-effective and agile but can lack strategic direction.

This system also refers to the complex network of actors, processes, and institutions involved in the creation and development of new technologies. It's a dynamic ecosystem with various components:

Academic Research: Universities, research institutions, and scientists play a key role in generating fundamental knowledge and laying the groundwork for future innovations.

Private Sector R&D: Companies invest heavily in research and development (R&D) to create new products and services that meet market needs and generate profits.

Government Funding: Government agencies often provide funding for research, particularly in areas like healthcare, energy, and defense, with the goal of advancing public interest.

Entrepreneurs and Inventors: Individual inventors and entrepreneurs can bring disruptive ideas to life, often through startups and small businesses.

Collaboration and Partnerships: Collaborations between different actors, like universities and businesses, can accelerate innovation by combining resources and expertise.

Technology Life Cycle

Once a technology is born, it embarks on a journey through various stages, each with its unique dynamics:

Innovation Stage: This early stage is marked by the conception and development of the technology. It involves significant research, prototyping, and experimentation, often with high levels of uncertainty and risk.

Growth Stage: If the technology finds traction, it enters a period of rapid adoption and market expansion. This stage is characterized by increasing profits, investments, and competition.

Maturity Stage: As the market becomes saturated, the growth slows down. This stage is marked by intense competition, cost optimization, and product differentiation.

Decline Stage: Eventually, the technology is overtaken by newer and more efficient alternatives. This stage witnesses declining demand, shrinking profits, and eventual obsolescence.

Understanding the technology life cycle is crucial for businesses and individuals alike. It allows for:

Strategic decision-making: Knowing where a technology is in its life cycle can help in making informed decisions about investment, product development, and market entry.

Identifying opportunities: The decline of technology can create opportunities for disruptive innovation, while the growth stage can offer lucrative market entry points.

Managing risk: Understanding the potential for obsolescence can help mitigate risks associated with investing in specific technologies.

Basics of Technology Transfer and Commercialization

Technology transfer and Technology Commercialization

Both technology transfer and technology commercialization play crucial roles in bringing groundbreaking innovation from the lab to the market, but they represent distinct steps in the journey. Here's a breakdown to highlight their key differences:

Technology Transfer

Focus: Moving technology from its creators (inventors, researchers) to another entity that will develop and utilize it. This entity can be a company, another research institution, or even a government agency.

Process: Often formal, involving legal agreements like licensing deals, patents, and other intellectual property protection measures. Technology transfer offices within universities and research institutions often facilitate this process.

Outcome: The focus is on making the technology available to the recipient, not necessarily on immediate commercialization. The recipient then takes responsibility for further development, production, and market launch.

Example: A university scientist invents a new medical device. They transfer the technology to a medical device company through a licensing agreement.

Technology Commercialization

Focus: Turning a technology into a profitable product or service for the market. This involves understanding market needs, designing a viable business model, and ensuring successful deployment and adoption.

Process: More flexible and entrepreneurial, often involving risk-taking and iterative development based on market feedback. Startups and established companies frequently undertake commercialization efforts.

Outcome: The focus is on generating revenue and creating value for both the technology developers and the consumers. It involves marketing, sales, and ongoing product improvement. Example: The medical device company from the previous example develops the device prototype, secures funding, conducts clinical trials, and launches the product in the market.

Here's an analogy to further illustrate the distinction:

Think of technology transfer as handing over the keys to a car. The new owner (recipient) can do whatever they want with it – keep it in the garage, take it for a spin, or even modify it.

Technology commercialization is like taking that car and opening a driving school. It focuses on making the car useful and profitable by creating a business around it.

In summary: Technology transfer bridges the gap between invention and development. Technology commercialization bridges the gap between development and the market.

Both are crucial elements in the innovation ecosystem, driving progress and economic growth by turning ideas into tangible solutions that benefit society.

Technology Commercialization Process

Technology commercialization is the exciting journey of converting creative ideas into practical products and services that solve real-world problems and generate value. It's a complex dance between science, business, and market forces, but understanding its core elements, models, systems, and processes can illuminate its path.

Applied or basic research generates a new technology or product
↓
Recognition of petential commercial applicability
↓
Invention Information Form
↓
Confidentiality Agreement
↓
Disclosure
↓
Technology Assessment
↓
Agency/Relationship Agreement
↓
Intellectual Propery Protection
↓
Commercialization Strategy
↓
Post-Agreement Management

Elements

Technology: The heart of the process, the innovation driving the commercialization efforts. It can be a new invention, a novel application of existing technology, or a significant improvement on a current solution.

Market: The target audience for the technology, defined by specific needs and challenges. Understanding the market landscape, trends, and competition is crucial for tailoring the commercialization strategy.

Commercialization Team: The skilled individuals who translate the technology into a viable business proposition. This team may include engineers, entrepreneurs, marketers, financiers, and legal experts.

Resources: The financial and physical resources needed to develop, produce, and market the technology. This includes funding, infrastructure, equipment, and intellectual property protection.

Models

Startup Model: A common approach for disruptive technologies, where a new company is formed to develop and introduce the technology to the marketplace. This model requires significant investment and entrepreneurial skills.

Licensing Model: Existing companies can license the technology from the inventors or research institutions, leveraging their established infrastructure and market reach. This can be a quicker route to market but may involve sharing profits and control.

Joint Venture Model: A collaboration between the technology owners and a strategic partner combines resources and expertise to commercialize the technology. This can be beneficial for complex technologies or those requiring significant investment.

Goldsmith Technology Commercialization Model

	Technical Market	**Market**	**Business**
Concept Phase			
Stage 1 Investigation	Step 1 Technology Analysis	Step 2 Market Needs Assessment	Step 3 Venture Assessment
Development Phase			
Stage 2 Feasibility	Step 4 Technical Feasibility	Step 5 Market Study	Step 6 Economic Feasibility
Stage 3 Development	Step 7 Engineering Prototype	Step 8 Strategic Market Plan	Step 9 Strategic Business Plan
Stage 4 Introduction	Step 10 Business Start-Up	Step 11 Pre-Production Prototype	Step 12 Market Validation
Growth Phase			
Stage 5 Growth	Step 13 Production	Step 14 Sales and Distribution	Step 15 Business Growth
Stage 6 Maturity	Step 16 Production Support	Step 17 Market Diversification	Step 18 Business Maturity

Important Technologies Commercialized by ICAR Crops and Horticulture

1. Bio-formulations

This category includes innovative biological products such as Arka Fermented Coco Peat, the Bio-bactericide Composition B5, and the DOR Bt-1 formulation. It also features diagnostic tools like a PCR-based detection kit for pomegranate diseases and a rapid detection system for Bt-Cry toxins. Additionally, bio-pesticides like *Paecilomyces lilacinus* and *Pochonia chlamydosporia* are used for pest management.

2. Post harvest Products and Technologies

Developments in this area include value- added products such as pickles from banana flowers and stems, fruit candies, ready-to- serve juices, and nutraceutical concentrates. Notable innovations include coconut chips, PUSA-branded snacks like Bajra Puff, Fruit Drink, Nutri Cookies, and Pearl Puf. Technological solutions include insulated and ventilated containers for transporting horticultural produce, an automatic pulp scooping machine, and lac-based coatings to extend the shelf life of fruits.

3. Crop protection Equipment

This group includes specialized tools and devices for pest control and insect rearing, such as insect rearing cages, aerial insect traps, UV chambers for *Corcyra* eggs, egg cleaning devices, pheromone and light traps, Helicoverpa oviposition cages, and other insect handling equipment.

Biotechnological Innovations

This includes genetic and molecular tools like the *Bacillus thuringiensis* Cry2Aa gene and simple sequence repeat (SSR) markers developed for Okra, aiding in crop improvement and research.

4. Improved Plant Varieties

A wide range of improved cultivars have been developed, including:

- **Wheat**: HI 1544
- **Rice**: Pusa RH-10, DRRH-2 and 3, Ajay (CRHR-7), CR Dhan 701 (CRHR-32), Rajalaxmi (CRHR-5)
- **Maize**: Vivek QPM 9, VL Babycorn-1
- **Chilli**: MSH 206, Arka Harita, Arka Meghana
- **Chrysanthemum**: Pusa Anmol
- **Cymbidium Orchids**: Hybrid varieties
- **Tomato**: Arka Rakshak
- **Turmeric**: IISR Alleppey Supreme, IISR Pratibha
- **Ginger**: IISR Varada
- **Nutmeg**: IISR Viswashree
- **Spices**: Ajmer Coriander-1, Ajmer Fenugreek-1 & 2, Ajmer Fennel-1

Systems

Technology Transfer Offices: Universities and research institutions typically establish dedicated Technology Transfer Offices (TTOs) to assist in moving their innovations to the private sector. These offices offer support in managing intellectual property, securing licenses, and identifying suitable partners for commercialization.

Incubators and Accelerators: These organizations provide early-stage startups with mentorship, funding, and networking opportunities to help them develop and commercialize their technologies.

Government Grants and Incentives: Many governments offer grants and other financial incentives to support technology commercialization, especially in areas of national priority.

Processes

Idea Validation: Thoroughly assess the market potential of the technology and identify the most promising application.

Technology Development: Refine the technology, ensuring its functionality, reliability, and manufacturability.

Business Plan Development: Create a comprehensive plan outlining the market strategy, financial projections, and team composition.

Funding and Partnerships: Secure the necessary funding and establish partnerships with key stakeholders.

Product Development and Manufacturing: Design, prototype, and manufacture the product or service.

Marketing and Sales: Launch a targeted marketing campaign and establish distribution channels.

Technology Transfer Model

The technology transfer model is a framework outlining the journey of an innovation from its conception in research institutions to practical application in the commercial world. The technology transfer process is often non-linear, involving possible revisions and refinements at various stages. Successful transfer relies heavily on clear communication and strong collaboration among researchers, technology transfer offices, and industry partners. Key challenges can include managing intellectual property rights, obtaining necessary funding, and addressing technical obstacles. Governments and public policies can play a role in incentivizing and supporting technology transfer, often through grants, tax breaks, and innovation ecosystems. It's a multi- stage process involving various stakeholders and activities:

Stage 1: Research

Discovery and invention: Researchers conduct fundamental or applied research, leading to breakthroughs or novel solutions with potential commercial applications.

Validation and proof-of-concept: Initial experiments and tests are conducted to assess the feasibility and effectiveness of the technology.

Documentation and intellectual property (IP) protection: The invention is documented, and potential IP rights, like patents or copyrights, are identified and secured.

Stage 2: Disclosure

Reporting and invention disclosure: The researchers or inventors report their findings to a dedicated technology transfer office (TTO) within their institution, often university or research institute.

Evaluation and assessment: The TTO evaluates the commercial potential of the technology, considering factors like market need, technical feasibility, and IP strength.

Prioritization and decision-making: Based on the evaluation, the Technology Transfer Office (TTO) determines whether to move forward with the technology's development and explore commercialization opportunities through potential partners.

Stage 3: Development

Prototype development: If chosen for further development, the technology is refined and transformed into a tangible prototype or proof-of-concept product.

Market research and analysis: Detailed market research is conducted to define the target audience, assess competition, and identify potential market entry strategies.

Business model development: A viable business model is created, outlining revenue streams, cost structures, and marketing channels.

Stage 4: Commercialization

Partnerships and licensing: Officers can explore collaborations with private firms or entrepreneurs who possess the necessary expertise and resources to commercialize the technology. This may include entering into licensing agreements, allowing the technology owner to earn royalties from its application.

Startup formation: In some cases, the technology may be commercialized through a new startup company founded by the inventors or other stakeholders.

Marketing and sales: Once the product or service is ready for market, the chosen partner or startup launches a marketing and sales campaign to reach the target audience.

Financial resources and infrastructure: Adequate funding and access to necessary infrastructure are vital for development and market launch.

Understanding the technology transfer model is crucial for researchers, entrepreneurs, policymakers, and anyone involved in bridging the gap between scientific breakthroughs and real-world solutions. By fostering a collaborative and supportive environment, we can accelerate the flow of innovative technologies from labs to markets, benefitting society as a whole.

Examples

A university technology transfer office licenses a new medical device technology to a private company for further development and commercialization.

A government agency funds a research project on renewable energy, with the expectation that the resulting technologies will be transferred to the private sector for implementation.

An inventor discloses their revolutionary new material to a technology accelerator, seeking partners and resources to bring it to market.

Different models can be used within the technology transfer framework:

Push model: Technologies are identified by researchers and then pushed towards potential commercial partners.

Pull model: Market needs are identified first, and then technologies are sought to address those needs.

Hybrid model: A combination of push and pull approaches is often used to maximize the chances of successful commercialization.

Key Factors for Success

Effective communication and collaboration: Clear communication between researchers, TTOs, and potential commercialization partners is crucial for success.

Strong intellectual property protection: Secure IP rights protect the technology from unauthorized use and incentivize investment.

Market-driven approach: Understanding market needs and tailoring the technology accordingly is essential for successful commercialization.

Technology Transfer Process

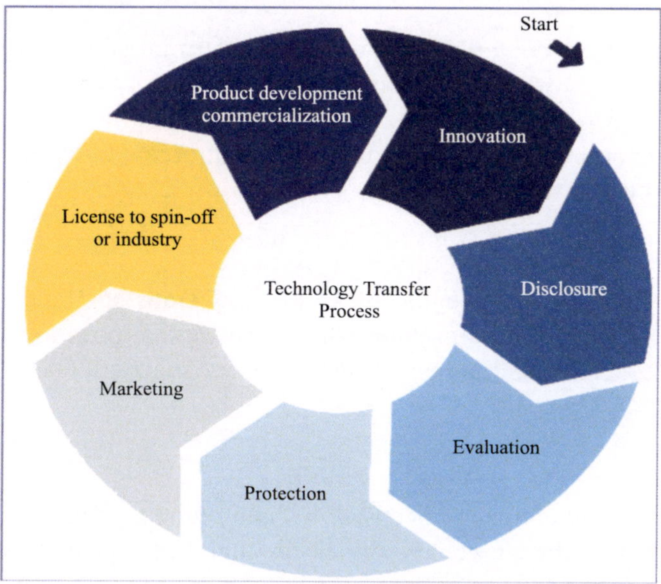

Research

Research activities frequently result in inventions through observations and experimentation. It's important to keep good, reliable laboratory records of your research activities for several reasons. Remember, to contact the Technology Transfer Office before you send or receive any research materials outside the University.

Pre-Disclosure

The inventor(s) should contact us early to discuss the invention, in order to determine if the timing is appropriate to submit an invention disclosure form.

Invention Disclosure

The inventor(s) must fill out and submit an Invention Disclosure Form to the Technology

Transfer Office, which marks the official start of the technology transfer process. Assessment

Invention Disclosure Forms received by the Technology Transfer Office are logged and evaluated. The technology is evaluated with the input of the inventor(s). Any necessary patent searches are conducted and market research is conducted to confirm the invention's commercialization potential.

Protection

Based on the assessment, the University will determine if it will file for patent protection on the invention. The inventor(s) should provide the attorney with any details that make the invention novel, useful, and non-obvious and any other information that may be required by the attorney, to obtain meaningful patent protection.

Marketing

In collaboration with the inventor, the Technology Transfer Office will carry out market research and pinpoint potential companies that possess the necessary expertise, resources, and industry connections to successfully commercialize the invention.

Licensing

The rights to an invention are licensed, without relinquishing ownership. In an exclusive license, the University will ensure that it retains the right to use and make the invention for research purposes so the inventor may continue his/her research.

License Management

The Technology Transfer Office will maintain and catalog the license for the life of patent. The performance of a license will be reviewed periodically and monitored to assure reimbursement for patenting costs and payments owed under the license.

Product Development

Developing a product primarily entails product design, engineering and testing. The start-up will also need to perform market research and market analysis to determine how the company plans to enter the market.

2

Intellectual Property Resource (IPR) Management Overview and Importance

Intellectual property (IP) encompasses creations of the human intellect, including inventions, artistic and literary works, as well as symbols, names, and designs. Intellectual property rights (IPR) are legal mechanisms that provide creators with exclusive rights to use, manage, and profit from their creations. These rights help protect the interests of innovators and promote continued creativity and innovation.

Intellectual Property Rights refer to the legal entitlements granted to individuals or organizations over their creative and intellectual works, such as patents, trademarks, and copyrights. These rights grant exclusive control or usage privileges for a specific duration, often creating a temporary monopoly. While the concept of intellectual property began to take shape in the 19th century, it was only during the 20th century that it was formally integrated into global legal frameworks.

Intellectual Property Rights (IPR) safeguard the creations of the human mind, such as inventions, artistic and literary works, brand names, logos, designs, and images used in commerce. These rights help foster innovation by allowing creators and inventors to benefit economically from their work.

Key Points

1. Intellectual Property Rights are legal protections acquired by creators or owners of intellectual works.
2. Intellectual property encompasses intangible assets that result from human creativity and thought.
3. These creations may include:
   - Scientific inventions
   - Artistic and literary works
   - Commercial designs, logos, and trademarks
   - Brand names and images

4. The primary objective of IP law is to:
   - Encourage the creation of a wide range of intellectual works.
   - Ensure a fair balance between protecting creators' rights and serving the public interest.
5. Article 27 of the Universal Declaration of Human Rights affirms that every individual has the right to enjoy the safeguarding of both moral and financial interests related to their scientific, literary, or artistic contributions.

IPR plays a Vital Role

Boosting innovation and creativity: By providing legal protection and financial rewards, IPR encourages individuals and companies to invest in research and development, leading to new discoveries and advancements.

Promoting fair competition: IPR ensures a level playing field for businesses by preventing unfair copying and imitation, allowing them to compete based on their own investment and innovation. Facilitating technology transfer and trade: Effective IPR systems create a secure environment for sharing and commercializing technology, promoting international trade and economic growth.

Protecting cultural heritage: IPR safeguards traditional knowledge, artistic expressions, and geographical indications, preserving cultural diversity and identity. Protecting IPR is crucial for:

Innovation: It encourages creators to invest in research and development by assuring them that they will benefit from the outcomes of their work.

Economic growth: Strong IPR frameworks attract foreign investment and stimulate knowledge- based industries.

Fair competition: IPR ensures that creators are not unfairly exploited by copycats and can compete on a level playing field.

Cultural diversity: It protects traditional knowledge and expressions of artistic creativity, fostering cultural diversity and identity.

Genesis of IPR

The concept of IPR has existed for centuries, evolving alongside human civilization. Early examples include guilds protecting the secrets of their trades and governments granting monopolies to inventors. The contemporary intellectual property rights (IPR) system took shape in the 19th century, driven by industrialization and the expansion of global trade. However, the idea of protecting intellectual creations has much older roots, with early versions of such protection existing in ancient Greece and Rome. However, the modern IPR system emerged in the 18th and 19th centuries with the Industrial Revolution, driven by the need to protect inventions and encourage technological advancements. International treaties such as the Berne Convention (1886) for safeguarding literary and artistic creations, and the Paris Convention (1883) for the protection of industrial property, played a key role in establishing a strong global foundation for intellectual property rights (IPR) protection.

IPR in India and Abroad

India has a well-established IPR system with comprehensive laws and regulations covering various forms of intellectual property. The Indian Patent Act, Copyright Act, Trademarks Act, Designs Act, Geographical Indications Act, and Semiconductor Integrated Circuits Layout- Design Act provide protection for various types of intellectual property. India is also a party to various international intellectual property rights (IPR) agreements, such as the Trade-Related Aspects of Intellectual Property Rights (TRIPS) agreement. However, India faces challenges in effectively enforcing its IPR laws, particularly in areas like software piracy and counterfeit goods. Addressing these challenges is key for nurturing a vibrant innovation ecosystem and attracting foreign investment. IPR systems vary significantly across different countries, with differences in the scope of protection, duration of rights, and enforcement mechanisms. Grasping these differences is crucial for individuals and businesses engaged in the global market. Most countries have national laws and international treaties to protect IPR.

Types of IPR

Patents: Protect inventions for a limited period (20 years in India) in exchange for disclosure of the invention to the public. This covers inventions like new products, processes, and devices.

1. A patent is a legal right given exclusively for an invention, which must be a novel product or process that fulfills specific criteria.

    1. Novelty,

    2. Non-obviousness, &

    3. Industrial use.

2. A patent grants the owner the authority to determine how others may use the invention or to prevent them from using it altogether.

Criteria for Issuing Patents in India

- **Novelty:** it should be new (not published earlier + no prior Public Knowledge/ Public Use in India)
- **Non-obviousness:** It should include an inventive element, meaning it must represent a technical advancement over existing knowledge and must not be obvious to someone skilled in the relevant area of technology.
- **Industrial use:** It should be capable of Industrial application

1. Patents in India are governed by "The patent Act 1970" which was amended in 2006 to make it compliant with TRIPS.

What Cannot be Patented?

1. **Frivolous Invention**: An invention that is harmful to public order, morality, or the health and well-being of humans, animals, or plants.
2. **Agricultural or Horticultural Methods**: Techniques or processes specifically related to farming or gardening practices.
3. **Traditional Knowledge**: Indigenous or ancestral wisdom and practices passed down through generations.
4. **Computer Program**: Software applications or code in isolation, without any technical application or hardware integration.
5. **Inventions Concerning Atomic Energy**: Discoveries or technologies associated with nuclear energy that fall under regulatory restrictions.
6. **Plants and Animals**: Biological varieties or species of plants and animals that are not eligible for patent protection.
7. **Scientific Principle Discovery**: The mere uncovering of a natural law or scientific concept without practical application.

Patent (Amendment) Rules, 2020

1. The central government has published an amended Patent (Amendment) Rules, 2020.
2. The new rules have **amended the format of a disclosure statement** that **patentees & licensees are required to annually submit to the Patent Office**.
3. The format contains disclosing the extent to which they have commercially worked or made the **patented inventions available to the public.**

4. The information must be submitted using Form 27, as mandated by the Patent Rules, 2003.
5. This legal obligation has been consistently overlooked by both patentees, licensees, and the Patent Office.
6. Multinational corporations and the United States have exerted considerable pressure to eliminate this requirement.

Criticism of Patent (Amendment) Rules, 2020

1. The amendment has significantly weakened the requirement of submitting information in the disclosure.
2. This could **hamper the effectiveness of India's compulsory licensing regime** which **depends on full disclosure of patent working information**.
3. This in turn could hinder access to vital inventions including life-saving medicines.

Copyrights: Protect original literary, artistic, and musical works from unauthorized copying or reproduction for a longer duration (life of the author plus 70 years in India). This includes books, music, films, paintings, and sculptures.

1. Copyright refers to the legal rights granted to creators for the protection of their original literary and artistic creations.
2. The scope of copyright includes a wide variety of works such as literature, music, artwork, films, software, databases, advertisements, maps, and technical illustrations.
3. **Copyrights in India are governed by "The Copyright Act, 1957" which was amended in 2012.**

Trademarks

1. To differentiate the products and services of one organization from another, elements such as logos, brand names, and taglines are used.
2. A trademark is a symbol or indicator that helps identify the goods or services of a particular business and sets them apart from those of other businesses.
3. The concept of trademarks has existed since ancient times, when craftsmen would inscribe their names or unique marks on items they created.

4. **Trademarks in India are Governed by Trade Marks Act 1999 which was Amended in 2010.**

Trade secrets: Confidential data that offers a business a competitive edge—such as formulas, production methods, or client databases is considered a trade secret.

- Trade secrets are a form of intellectual property (IP) that protect sensitive information, which can be commercially transferred or licensed.
- Obtaining, using, or revealing this confidential information without permission is considered an unfair practice and a breach of trade secret laws.
- There is no specific law.

Geographical indications: Protect the names and reputations of products originating from a specific geographical location. Examples include Darjeeling tea and Champagne.

- A GI tag is a legal recognition given primarily to an agricultural, natural or a manufactured product (handicrafts & industrial goods) originating from a definite geographical territory.
- A Geographical Indication (GI) serves as a certification that a product possesses specific qualities and a unique identity, primarily due to its geographical origin.
- Typically, a geographical indication features the name of the region or location where the product is produced.
- Once the GI protection is granted, **no other producer can misuse the name** to market similar products.
- It also provides comfort to customers about the authenticity of that product.
- **Geographical Indicators in India are governed by "The Geographical Indications of Goods (Registration & Protection) Act, 1999"**.

Industrial designs: Protect the ornamental or aesthetic aspects of an article/object. This includes the shape, configuration, pattern, or colour of a product.

Industrial Designs in India are governed by "The Designs Act 2000".

Emergence of IPR Regimes and Governance Frameworks

Several international treaties and organizations have been established to harmonize and strengthen IPR protection globally:

- The Trade-Related Aspects of Intellectual Property Rights (TRIPS) Agreement, established under the World Trade Organization (WTO), outlines the minimum requirements that member countries must follow for protecting intellectual property rights (IPRs) within their national legal systems.
- TRIPS is an international agreement on intellectual property rights.
- It is a landmark and most comprehensive treaty on Intellectual property.
- While earlier treaties' subject matters were specific, TRIPS deal with 8 kinds of property rights –
  - Patents,
  - Trademarks,
  - Trade Secrets,
  - Copyrights,
  - Industrial Designs,
  - Plant Varieties,
  - Integrated Circuits and layouts, and
  - Geographical Indication.
- Further, almost all countries are party to TRIPS. In earlier treaties, only limited countries participated.
- It also establishes a system for enforcement that was not included in the WIPO agreements.
- It mandated all member countries to make their domestic laws complaint to TRIPS.
- India passed certain laws and amended others. India's IPR regime now stands fully compliant to TRIPS.
  - For example, India amended patent law in 2005 to provide 'product' patent protection. Earlier protection was available only to 'processes'.
- TRIPS was the outcome of discussions held in the Uruguay round which directed to the formation of WTO. This treaty is an offshoot of the General Agreement on Trade in Goods (GATT). This treaty provided a robust Dispute Resolution Mechanism and stringent penal provisions under the auspices of WTO.

- It was implemented in 1995 and is mandatory for all World Trade Organization (WTO) member countries.

Convention on Biological Diversity (CBD)

Aims to promote the fair and just distribution of benefits derived from the use of genetic resources. It acknowledges the value of preserving traditional knowledge and the genetic heritage of indigenous communities. The agreement also urges nations to create unique, tailor- made (sui generis) systems to safeguard these types of intellectual property.

The **Cartagena Protocol on Biosafety**: a supplementary agreement to the CBD, addresses the safe handling and international transfer of living modified organisms (LMOs). It is primarily concerned with ensuring biosafety and protecting biodiversity, especially in relation to advancements in modern biotechnology.

Basics and Backgrounds

1. The United Nations Convention on Biological Diversity (UNCBD) is an international treaty dedicated to preserving global biological diversity.
2. Commonly referred to as the Biodiversity Convention, this multilateral treaty was opened for signature during the 1992 Earth Summit held in Rio de Janeiro. It plays a vital role in promoting sustainable development and operates under the United Nations Environment Programme (UNEP).
3. A total of 196 countries are signatories to the Convention on Biological Diversity (CBD).
4. India is among the countries that have ratified the Convention.
5. The treaty is legally binding for all countries that have signed and ratified it.
6. The Convention is governed by the Conference of the Parties (COP), which comprises the governments of the ratifying nations.
7. The CBD's Secretariat is headquartered in Montreal, Canada.
8. Out of all UN member states, only the United States and the Vatican have not become parties to the Convention.
9. At the 1992 Earth Summit, two major legally binding agreements were adopted: the UNCBD and the United Nations Framework Convention on Climate Change (UNFCCC).

10. Initially, over 150 countries signed the Biodiversity Convention at the Summit, and more than 175 have since ratified it.

Objective

- The preservation of biological diversity, the sustainable utilization of its elements, and the fair and equitable distribution of benefits derived from the use of genetic resources—along with the appropriate transfer of relevant technologies—while respecting all associated rights and ensuring adequate financial support.

Goals

1. Preservation of biological diversity
2. Responsible and long-term use of biodiversity components
3. Just and equal distribution of benefits derived from genetic resources

To achieve the above mention goals CBD follow certain Targets and Protocols as follows

Goals	Targets/Protocols
Protect biodiversity	COP meetings, Aichi Targets
Safe use of bio-technology	Cartagena Biosafety Protocol
Stop unfair use of Genetic resources	Nagoya Genetic Resources Protocol

Functions

1. **Recognizing the Inherent Worth of Biodiversity:** Acknowledging that biodiversity holds intrinsic value beyond its utility.
2. **Viewing Biodiversity Conservation as a Shared Global Responsibility:** Emphasizing that preserving biodiversity is a collective concern of all people.
3. **State Responsibility for Conservation and Sustainable Use:** Committing the state to conserve biodiversity and ensure its sustainable use.
4. **Acknowledging State Sovereignty Over Biological Resources:** Affirming the sovereign rights of the state over its biological resources.
5. **Adopting a Precautionary Approach:** Encouraging proactive measures to prevent biodiversity loss, even in the face of scientific uncertainty.
6. **Recognizing the Role of Local Communities and Women:** Highlighting the critical contributions of local populations and women in conserving biodiversity.

7. **Promoting Technology Access and Financial Support:** Supporting the availability of technologies for developing nations and advocating for new and additional financial resources to combat biodiversity decline in the region.

International Union for the Protection of New Varieties of Plants (UPOV)

- UPOV is an intergovernmental body based in Geneva, Switzerland, focused on encouraging the innovation and protection of new plant varieties.
- It was founded through the International Convention for the Protection of New Varieties of Plants, initially adopted in Paris in 1961 and later revised in 1972, 1978, and 1991.
- UPOV's core mission is to establish an effective plant variety protection system to promote plant breeding, ultimately benefiting society.
- Under the UPOV Convention, member countries grant **breeder's rights**, which are a form of intellectual property given to developers of new plant varieties.
- Breeders must give permission for the commercial propagation of their protected varieties.
- Protection under UPOV can only be obtained by the original breeder of the plant variety.
- The definition of a breeder is broad and includes individuals, farmers, researchers, public institutions, and private companies.
- India is not currently a member of UPOV.

Bay of Bengal Initiative for Multi-Sectoral Technical and Economic Cooperation (BIMSTEC)

BIMSTEC is a regional multilateral organization that promotes cooperation among countries in South Asia and Southeast Asia. It includes a dedicated working group on Intellectual Property Rights (IPR), which aims to harmonize IPR-related laws and frameworks across the member nations.

What is Bimstec?

1. The Bay of Bengal Initiative for Multi-Sectoral Technical and Economic Cooperation is a regional organization that brings together countries located around the Bay of Bengal.

2. It comprises nations forming a geographically contiguous area.
3. Among the seven member countries: Five belong to South Asia:
   - Bangladesh
   - Bhutan
   - India
   - Nepal
   - Sri Lanka

   Two are from Southeast Asia:
   - Myanmar
   - Thailand
4. BIMSTEC serves as a bridge between South and Southeast Asia and links the ecological systems of the Great Himalayas with those of the Bay of Bengal.
5. The initiative is primarily focused on fostering economic growth, social advancement, and cooperation on shared regional concerns.

Origin of BIMSTEC

1. The organization was established in 1997 with the adoption of the Bangkok Declaration.
2. It began with four founding members — Bangladesh, India, Sri Lanka, and Thailand — under the acronym 'BIST-EC'.
3. After Myanmar joined later in 1997, it became 'BIMST-EC'.
4. In 2004, the inclusion of Nepal and Bhutan led to the current name — Bay of Bengal Initiative for Multi-Sectoral Technical and Economic Cooperation (BIMSTEC).

Core Objectives

1. To foster an environment conducive to rapid economic development in the region.
2. To uphold principles of equality and mutual partnership.
3. To encourage collaborative efforts and shared support in areas of mutual interest.

4. To promote cooperation in education, science, technology, and related fields.

Guiding Principles

1. Respect for Sovereign Equality
2. Preservation of Territorial Integrity
3. Assurance of Political Independence
4. Non-interference in Internal Matters
5. Peaceful Coexistence
6. Mutual Benefits
7. Complementary to, and not a replacement for, existing bilateral, regional, or multilateral cooperation among members

Organizational Structure

1. **BIMSTEC Summit:** The top decision-making body, attended by the heads of state or government.
2. **Ministerial Meeting:** The second-highest decision-making forum, attended by the foreign or external affairs ministers.
3. **Senior Officials' Meeting:** Comprises high-level officials from foreign ministries of member countries.
4. **BIMSTEC Working Group:** Includes ambassadors or appointed representatives of member states, meeting monthly at the Secretariat in Dhaka.
5. **Business and Economic Forums:** Platforms that engage the private sector to ensure their involvement in regional development.

Understanding these institutional structures is key to managing intellectual property rights effectively and ensuring robust global protection for innovations and creative works.

IPR Protection Systems and National Policy

India's National IPR Policy, launched in 2016, aims to nurture a strong innovation ecosystem. The policy emphasizes both the protection of intellectual property and the promotion of accessible knowledge and technology, thus balancing creativity with public interest.

1. The policy seeks to position intellectual property rights (IPRs) as valuable financial assets, fostering innovation and entrepreneurship while ensuring public interest is safeguarded.
2. The strategy will undergo evaluation every five years, involving consultations with relevant stakeholders.
3. To ensure robust and effective IPR legislation, the existing laws will be periodically reviewed, revised, and refined to eliminate discrepancies and improve effectiveness.
4. The policy fully aligns with the WTO's TRIPS agreement.
5. The Department for Promotion of Industry and Internal Trade (DPIIT) serves as the central authority handling all matters related to IPR.
6. The policy maintains key provisions such as Compulsory Licensing (CL) and restrictions against the practice of patent evergreening, as outlined in Section 3(d) of the Indian Patents Act.
7. As per the Indian Patents Act, a Compulsory License may be granted for a pharmaceutical product if it is considered unaffordable, allowing approved generic manufacturers to produce the medicine under specific conditions.

Objectives

Stimulate innovation and creativity: Encouraging research and development, fostering a culture of innovation, and rewarding creators through effective IP protection.

Promote economic growth and development: Attracting foreign investment, facilitating technology transfer, and creating high-skilled jobs through IP-driven industries.

Enhance India's global competitiveness: Strengthening India's position as a knowledge economy and a hub for innovation by aligning with international IPR standards.

Ensure equitable access to knowledge and technology: Ensuring a fair balance between protecting creators' rights and meeting users' needs by supporting access to cost-effective healthcare, education, and technology.

Need for an IPR Policy

1. The policy plays a crucial role in enabling the government to offer tax incentives, thereby promoting research and development activities.
2. This policy has been introduced in response to the U.S. Trade Representative (USTR) placing India on the 'Priority Watch List' in its 2016 Special 301 Report, which assesses global IPR protection and enforcement.
3. It aims to uphold the integrity of innovations, thereby reducing legal disputes in the intellectual property domain.
4. The policy will contribute to safeguarding India's rich heritage of traditional knowledge.
5. It supports the growth and success of initiatives like Make in India, Startup India, and Digital India by providing a stronger intellectual property framework.

Key Features

Focus on specific sectors: The policy prioritizes sectors like pharmaceuticals, biotechnology, electronics, and traditional knowledge, recognizing their potential for economic growth and social development.

Strengthening IP administration: Streamlining the process of filing and granting patents, trademarks, and copyrights, reducing pendency and improving efficiency.

Promoting awareness and education: Enhancing public awareness about the importance of IPRs and their role in fostering innovation, through education and outreach programs.

Supporting startups and SMEs: Providing access to funding, mentorship, and IP resources to encourage innovation and entrepreneurship among startups and small and medium sized enterprises (SMEs).

Protecting traditional knowledge: Recognizing and safeguarding the IPR of indigenous communities and traditional communities.

Drawbacks

1. India's IPR Policy reiterates its dedication to the Doha Declaration under the WTO's TRIPS Agreement, highlighting the importance of safeguarding public health. However, there are concerns that referencing this flexibility might lead to attempts to exploit TRIPS provisions, potentially benefiting pharmaceutical firms disproportionately.

2. Without proper government funding and supportive programs to ensure universal access to medicines, any legislative changes may negatively impact both the domestic generic drug industry and community health in India.

3. The policy does not address problems related to traditional knowledge or the informal innovations rooted in such knowledge systems.

4. There is little evidence suggesting that modern laws like utility models or trade secret protections effectively support grassroots and informal innovations.

5. The policy proposes significant public investment in protecting foreign intellectual property in India, even though it acknowledges that the primary responsibility for IP protection lies with the rights holders. Furthermore, the mention of state legislation in the context of copyright protection suggests that the policy favours IP owners over broader societal interests.

6. Some experts argue that the National IPR Policy is too vague and lacks the concrete strategies necessary to truly stimulate innovation.
7. The policy appears to equate a higher number of IP filings with greater innovation, overlooking the fact that intellectual property serves as a means to an end to achieve innovation, not the ultimate goal.
8. It promotes the idea that all knowledge should be transformed into intellectual property.
9. Even corporations have acknowledged that intellectual property frameworks are not always effective in certain technology fields, where open access to knowledge may be more beneficial.
10. Experts also question the policy's applicability to the rural informal sector, noting that the unique nature of rural innovation is poorly understood. Imposing a formal IP system in such settings may actually be detrimental.
11. The decision to criminalize violations under the Indian Cinematograph Act is seen as excessive and overly punitive.
12. Intellectual property infringements are fundamentally civil matters and should not be treated as criminal offenses.

Opportunities

While the National IPR Policy has made significant progress, challenges remain in its implementation:

Enhancing enforcement: Strengthening mechanisms to combat piracy and counterfeit goods to ensure effective protection of IPRs.

Addressing public concerns: Ensuring fair access to knowledge and technology while protecting the rights of creators and addressing the needs of users.

Building capacity: Equipping India's workforce with the necessary skills and knowledge to effectively manage and utilize intellectual property.

IPR Laws

IPR laws cover a broad and intricate legal framework, but they can be simplified by categorizing them according to the different types of intellectual property they protect.

1. Patents

Key laws: The Indian Patent Act, 1970, governs patentability, application processes, examination procedures, and infringement rules.

What is protected: Inventions new to the world, non-obvious, and capable of industrial application. Examples include chemical compounds, manufacturing processes, and mechanical devices.

Duration of protection: 20 years from the filing date.

2. Copyrights

Key laws: The Copyright Act, 1957, covers the protection of original literary, artistic, dramatic, musical, and cinematographic works.

What is protected: Original expressions of ideas, including books, music, films, paintings, sculptures, and software.

Duration of protection: Life of the author plus 70 years in India.

3. Trademarks

Key laws: The Trademarks Act, 1999, governs the registration, protection, and infringement of trademarks.

What is protected: Distinctive signs that identify the source of goods or services, such as logos, brand names, and slogans.

Duration of protection: Initially 10 years, renewable indefinitely upon payment of fees.

4. Trade Secrets

Key laws: Although not explicitly defined in a single statute, trade secrets are protected under common law principles and the Indian Contract Act, 1872.

What is protected: Confidential information that gives a business a competitive advantage, such as formulas, manufacturing processes, and customer lists.

Duration of protection: As long as the information remains confidential and provides a competitive advantage.

5. Geographical Indications (GIs)

Key laws: The Geographical Indications of Goods (Registration and Protection) Act, 1999, protects the names and reputations of products originating from a specific geographical location. What is protected: Products having unique qualities attributable to their geographical origin, such as Darjeeling tea and Alphonso mangoes.

Duration of protection: Perpetual.

6. Industrial Designs

Key laws: The Designs Act, 2000, governs the registration, protection, and infringement of industrial designs.

What is protected: The ornamental or aesthetic aspects of articles, such as the shape, configuration, pattern, or color of a product.

Duration of protection: 10 years, extendable for another 5 years. Additionally:

The Semiconductor Integrated Circuits Layout-Design Act, 2000, protects the unique layout of semiconductor chips.

The Biological Diversity Act, 2002, recognizes and protects traditional knowledge of indigenous communities.

Understanding these key laws and the types of intellectual property they protect is crucial for individuals and businesses to effectively utilize and navigate the IPR landscape. Remember, IPR laws are not static, evolving with technological advancements and international agreements.

Procedures for Filing IP protection

Filing for IP protection can seem daunting, but with the right information and guidance, it can be a smooth process. Here's a breakdown of the procedure for different types of intellectual property

1. Patents

Stage 1: Pre-filing search and drafting: Conduct a thorough patent search to ensure your invention is truly unique and novel. Then, draft a detailed patent application, including specifications, claims, and drawings, with the help of a patent attorney.

Stage 2: Filing: Submit your application to the Indian Patent Office (IPO) along with the prescribed fees. You can choose online or offline filing methods.

Stage 3: Examination: The IPO examiner will review your application to ensure it meets all patentability requirements. This may involve back-and-forth communication to address any objections.

Stage 4: Grant: If your application meets all requirements, the IPO will grant you a patent, giving you exclusive rights to exploit your invention for 20 years.

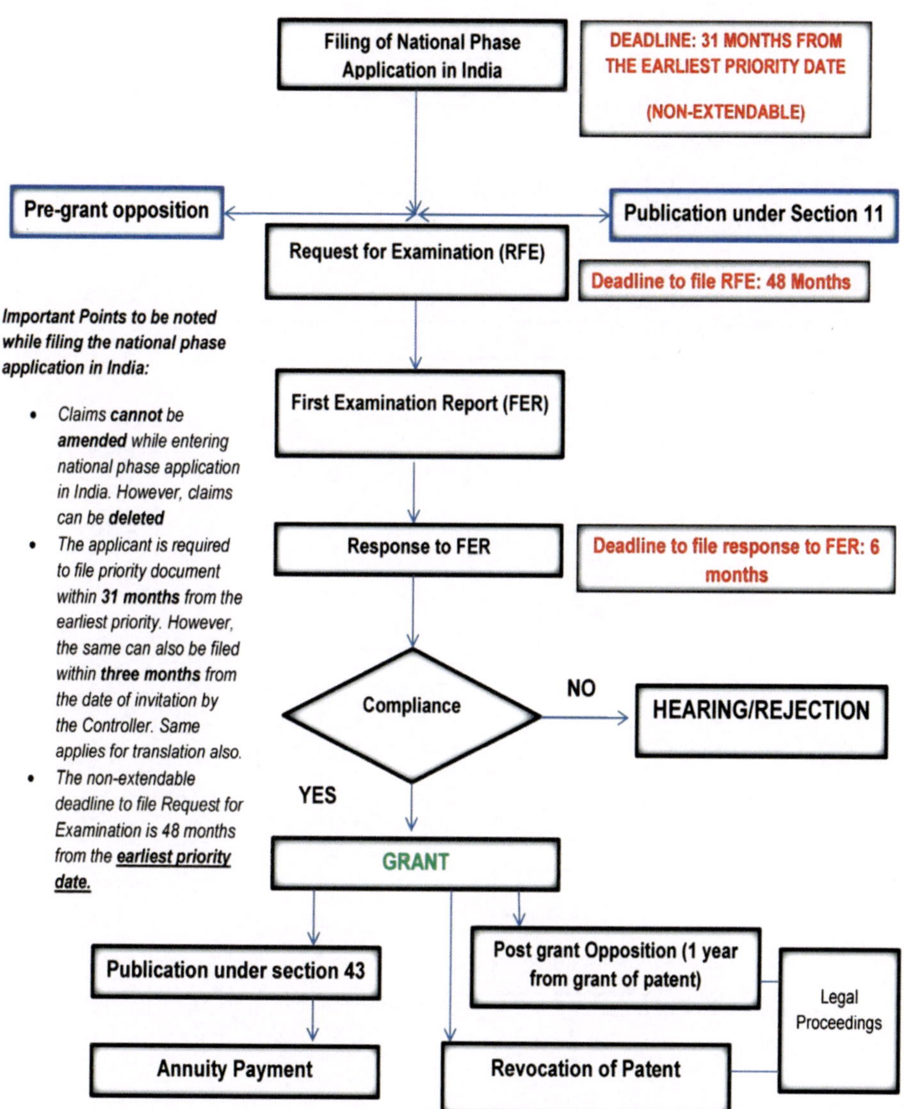

Formal Requirements:

- *Power of Attorney: 3 months from the date of filing the application in India*
- *Proof of Right: 6 months from the date of filing the application in India*

2. Copyrights

Stage 1: Creation and fixation: Your work is automatically protected from the moment it is created and fixed in a tangible medium (e.g., written on paper, recorded on digital media).

Stage 2 (Optional): Registration: While registration is not mandatory for copyright protection, it provides stronger legal evidence and simplifies infringement claims. You can register your work online through the Copyright Office website.

Stage 3: Enforcement: If your copyright is infringed, you can take legal action against the infringer to seek damages and injunctions.

Copyright Registration Workflow

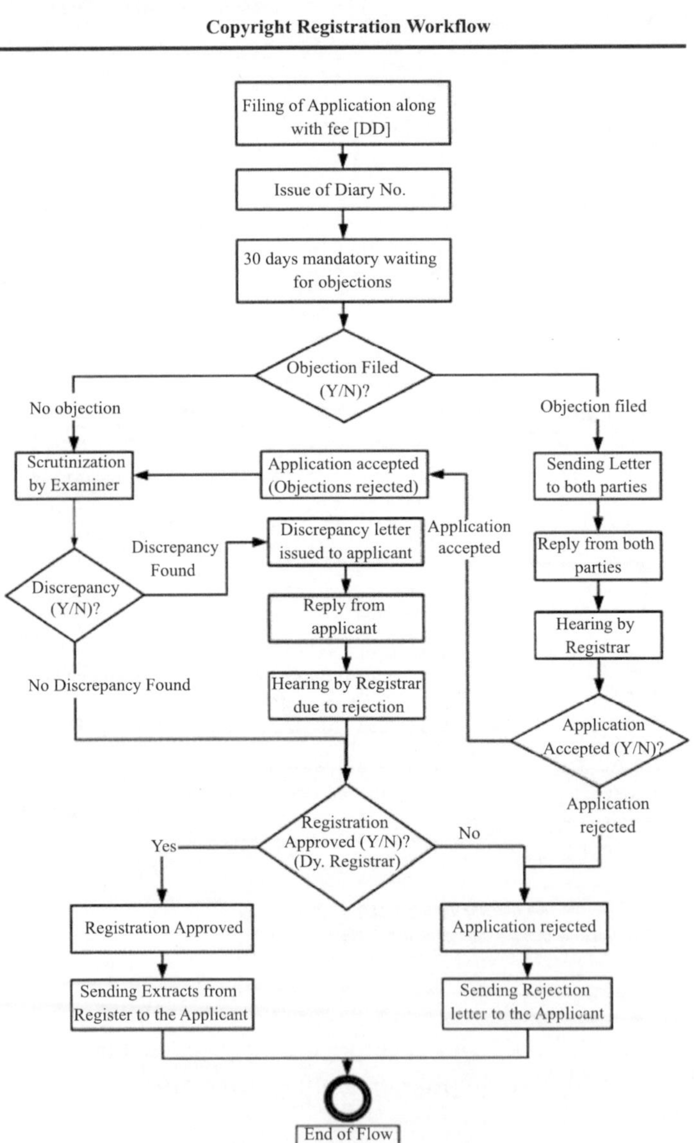

3. Trademarks

Stage 1: Trademark search: Conduct a thorough trademark search to ensure your chosen mark is not already registered by another company in your product/service category.

Stage 2: Application: File a trademark application with the Trade Marks Registry, specifying the trademark, product/service categories, and other details.

Stage 3: Examination and publication: The Registry will examine your application for compliance with trademark rules and, if approved, publish it in the Trademark Journal for opposition.

Stage 4: Registration: If no opposition is filed within the specified period, your trademark will be registered, granting you exclusive rights to use it for 10 years, renewable indefinitely.

Applications for registration of trademarks

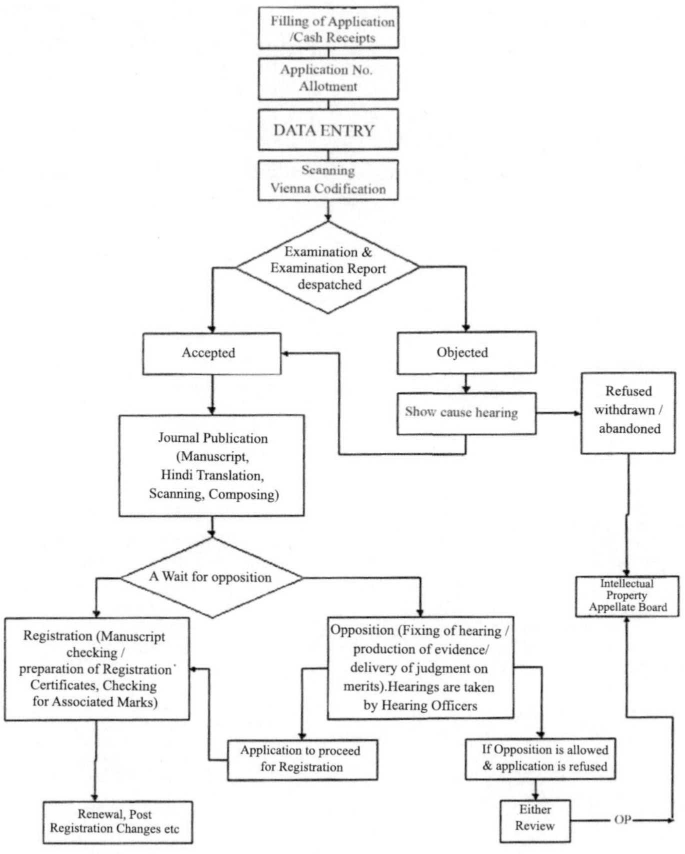

4. Trade Secrets

Stage 1: Identify and classify: Clearly identify your trade secrets and classify them by level of sensitivity and importance.

Stage 2: Implement protective measures: Take steps to maintain the confidentiality of your trade secrets, such as through non-disclosure agreements, access controls, and employee training.

Stage 3: Enforcement: If your trade secret is misappropriated, you can take legal action against the party responsible based on breach of contract or common law unfair competition principles.

5. Geographical Indications (GIs)

Stage 1: Application: File an application for GI registration with the Registrar of Geographical Indications, providing details about the product, its origin, unique qualities, and evidence of its geographical link.

Stage 2: Examination and opposition: The Registrar will examine the application and publish it for opposition from anyone who believes the GI claims are inaccurate or misleading.

Stage 3: Registration: If no valid opposition is filed, the GI will be registered, granting the producer community exclusive rights to use the GI for their product.

The registration process - flow chart

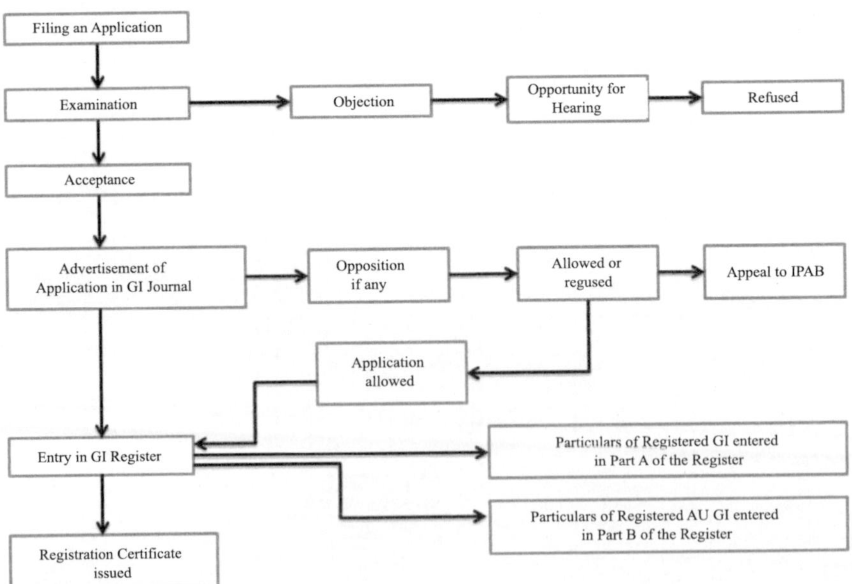

Protecting and Managing Intellectual Property in Agricultural Research: A Multi- Stakeholder Approach

Systems of IP Protection and Management in Agricultural Universities and Research Institutions

1. The IP rights accruing to ICAR in various forms would be embodied in the following

   Indian Acts, as amended from time to time

   a. The Copyright Act, 1957 as amended in 1983, 1984, 1992, 1994, 1999 2012 along with Rules 1958, 2013 and 2016

   b. The Patents Act, 1970 as amended in 1999, 2002, 2004 (Ordinance), 2005, along with Rules 1972, 2003, 2005, 2006, 2010 and 2016

   c. The Trade Marks Act, 1999 as amended in 2010 and 2013 along with Rules 1999, 2002, 2010, 2013 and 2017

   d. The Industrial Designs Act, 2000 along with Rules 2001, 2008 and 2014

   e. The Geographical Indications of Goods (Registration and Protection) Act, 1999 along with Rules 2002

   f. The Semiconductor Integrated Circuits Layout-Design Act, 2000 along with Rules 2001

   g. The Protection of Plant Varieties and Farmers' Rights Act, 2001 along with Rules 2003, 2006 and 2009

2. The Biological Diversity Act, 2002, along with its 2004 Rules, outlines the procedures for accessing biological and genetic resources for agricultural research and their protection under intellectual property rights.

3. Among the various forms of IPR under the respective acts, ICAR primarily focuses on securing patents, plant variety protections, copyrights, and trademarks. For trade secrets and other confidential information, protection is ensured through appropriate confidentiality agreements based on specific situations.

4. The Protection of Plant Varieties and Farmers' Rights (PPV&FR) Act aligns with Article 27.3(b) of the TRIPS Agreement. All varieties developed by ICAR that are notified under Section 5 of the Seeds Act, 1966, and have not exceeded 15 years from the date of notification, are eligible for registration and protection under this Act.

5. Copyrights, whether officially registered or not, exist for all creative works produced by ICAR scientists and institutional contributors. However, registering copyrights, especially for software, databases, and digital content, can offer more robust protection. Trademarks, collective marks, and industrial designs are also relevant for ICAR. While geographical indications have broader implications, they are not central to ICAR's core research. Layout designs of integrated circuits may be relevant to a few specific agricultural research areas.
6. National-level authorities have been constituted under the PPV&FR Act and the Biological Diversity Act to oversee functions and responsibilities assigned by these laws. ICAR respects the jurisdiction of these statutory bodies and ensures that its activities do not interfere with their mandates. Nevertheless, ICAR may collaborate in inter-departmental assessments, provide scientific inputs, and offer technical assistance where necessary for the benefit of Indian agriculture and farming communities. In the context of geographical indications, ICAR can play an advisory or facilitative role when required.
7. The management of ICAR's intellectual property and technology transfer activities will adhere to the existing national IPR legislation, policies, and guidelines. Scientists and institutional bodies within ICAR are expected to follow the framework and directives related to IPR management and commercialization. In complex or unclear situations, the relevant authority at ICAR headquarters will be consulted for specific guidance.
8. ICAR will pursue IP protection in compliance with Indian legal provisions and international agreements to which India is a signatory. The organization aims to promote the transfer of technologies backed by IPR—including products, processes, and innovations—through both commercial and public dissemination channels. Effective IP management will foster a commercially oriented research environment in the public sector, aiding in the transformation of Indian agriculture into a competitive, global-scale enterprise.

Agricultural research generates valuable knowledge and technologies with immense potential to improve food security, productivity, and sustainability. Protecting and managing intellectual property (IP) in this domain is crucial for several reasons:

Incentivizes research and development: By securing IP rights, researchers and institutions can gain financial rewards and recognition, spurring further advancements in agricultural technologies.

Facilitates technology transfer: Effective IP management enables efficient licensing and partnerships with private companies and other stakeholders, bringing innovations to farmers and the market faster.

Promotes global collaboration: Clear IP frameworks can foster international research collaborations and knowledge sharing, benefiting researchers and farmers across borders.

Systems and Processes

Agricultural universities and research institutions play a key role in developing and managing IP within this field. Some of the key systems and processes they implement include:

Technology disclosure and invention management: Establishing clear procedures for researchers to disclose inventions and manage the initial stages of IP protection.

IP policy and guidelines: Formulating a comprehensive IP policy outlining ownership, rights, and responsibilities regarding inventions and technologies developed within the institution.

Technology transfer offices: Dedicated offices that assist researchers in navigating the IP process, including patent filing, licensing, and commercialization strategies.

Partnerships and collaborations: Working with private companies, government agencies, and NGOs to facilitate technology transfer and leverage commercialization expertise.

Stakeholder Roles

Beyond research institutions, various stakeholders play crucial roles in the IP protection and management within agricultural research:

Government agencies: Provide funding for research, establish legal frameworks for IP protection, and incentivize technology transfer through grants and policies.

Private companies: Partner with research institutions for technology development, license and commercialize innovations, and drive market adoption of new technologies.

Farmers and farmer organizations: Provide feedback on research priorities, participate in trials and demonstrations, and benefit from the adoption of improved technologies.

NGOs and civil society organizations: Advocate for equitable access to knowledge and technologies, promote sustainable agricultural practices, and ensure environmental considerations are integrated into IP management.

Challenges and Opportunities

Effective IP protection and management in agricultural research faces several challenges: Limited resources: Research institutions often lack the expertise and funding to navigate the complex IP landscape.

Balancing incentives and access: Ensuring researchers benefit from their inventions while promoting affordable access to technologies for farmers in developing countries.

Biopiracy and unfair appropriation: Protecting traditional knowledge and genetic resources of indigenous communities from exploitation.

Despite these Challenges, Numerous Opportunities Exist

Leveraging digital technologies: Using online platforms and databases to improve access to IP information and facilitate technology transfer.

Promoting open-source models: Exploring alternative IP frameworks like open-source licenses to ensure broader access to valuable agricultural innovations.

Building capacity and awareness: Providing training and resources to researchers and stakeholders to understand and utilize IP effectively.

Protecting and managing IP in agricultural research is a complex but crucial endeavour. By adopting comprehensive systems, fostering multi-stakeholder collaboration, and addressing emerging challenges, we can ensure that innovation in this essential domain benefits not only researchers and institutions but also farmers, communities, and the global food system as a whole.

Management of IPR Mechanisms of IPR Management

Mechanisms of IPR Management – Institutional arrangement

Stage	*Mechanism*	*Description of the mechanism*	*Target of intellectual property rights*
1. Exploration	Intellectual property agreement	1. Contributions from each participant (form and value) 2. Statement of prior knowledge, technologies and knowhow to its value and property rights it confers. 3. Indemnity clauses	
		4. Confidentiality clauses	Technology and knowhow that belonging to: The own organization, your allies and other organizations
	Proceedings	1. Participants 2. Entry or exit of participants 3. Key considerations 4. Decisions 5. Intellectual property implications	
	Tracking records	1. Any product that can be protected by intellectual property rights 2. Contributions in prior knowledge, knowhow, methodologies, financial resources, human, etc. 3. Generation of new knowledge and creations 4. Participation of natural persons or companies at each stage of the project 5. Any measure to ensure the protection of in- tellectual property of intangible assets arising in the project	
2. Development	Agreements and /or confidentiality statements	Generally, all aspects of product development, their potential uses, impacts, scope and users could be protected by way of confidentiality agreements	Design testing feedback

Stage	***Mechanism***	***Description of the mechanism***	***Target of intellectual property rights***
	Cession of property rights	By developers and participants of the entity that will commercialize the technology or that will make the transfer process or that continue the development	
	Authorization for use of data	Personal data that can be used as inputs in the creative process	
	Authorization for use of photographic records, video and audio	They can be used as inputs in the creative process	
3. Implementation	Business models supported by IP. structuring technology packages	1. Patents 2. Utility models 3. Industrial designs, 4. Knowhow protected by trade secret, 5. Distinctive signs (brands, quality certificates, among others 6. Copyright	Commercialization transfer follow-up to the IP licensed or operated by self-employed
	Policies And Strategies		

Stage 1: Exploration

The objective is to identify opportunities to enhance existing products or services, or to develop entirely new ones. This involves employing specific methods such as co-creation strategies, technological surveillance, resource and capability assessments, feasibility evaluations (if applicable), and market research. These tools help in identifying intellectual property (IP) owners, users, customers, and strategic partners who will apply or benefit from the knowledge generated. During this phase, IP agreements are established to clearly define the roles, contributions, and ownership rights of all participants. These agreements also include detailed documentation of the technology and know-how involved, indemnity clauses, meeting records, activity logs, and other essential project documents. IP management can be approached through three key categories:

1. **Internal IP**, which includes all intellectual property rights and know-how owned by the organization. It is advisable to create an inventory of existing technologies, focusing on the knowledge chain elements related to those technologies.
2. **Partner IP**, which comprises intellectual property and know-how contributed by organizations collaborating on the innovation project. In this case, it is essential to document the terms of collaboration, how the technology will be integrated, and any licensing or usage conditions that apply.
3. **External IP**, referring to intellectual property from entities not directly involved in the collaboration. When incorporating such third-party technologies, it is crucial to obtain proper licenses to ensure lawful use and integration within the project.

Stage 2: Development

At this phase, leaders need to oversee the execution of non-disclosure agreements, confidentiality clauses, property rights transfers, and permissions for data use—including the use of photographs, videos, and audio recordings. This is essential because concepts, prototypes, and scaled evaluations may already exist, forming the basis for proposed business models related to the emerging goods and services (Kaiser, 2010). Intellectual Property Management (IPM) during this stage centers on the design, testing, and feedback activities within the co-creation process.

1. **IP Management in the Design Phase**: This phase involves the responsible use of existing knowledge and ensuring that the intellectual property rights of third parties are respected.
2. **IP Management in the Testing Phase**: During testing, it is essential to use standard agreements or sign non-disclosure agreements (NDAs) with external experts or institutions involved in the process. This helps protect sensitive information that, if revealed, could compromise the future IP protection of the innovation. Since co- creation with users or customers is involved, obtaining their consent for the use of photos, videos, audio, and personal data is critical, as these elements contribute to the creative development.
3. **IP Management in the Feedback Phase**: At this point, inputs are collected from experts, users, R&D teams, and collaborators to evaluate the outcomes and make informed decisions regarding the project's direction and the technologies used. Based on this feedback, the following actions may be considered:
    1. Initiating the protection of innovations through appropriate IP tools;
    2. Discontinuing certain technologies, which may require renegotiation if they are already under license;
    3. Adding new technologies that are protected by IPR, which would involve acquiring licenses or rights;
    4. Creating additional technologies to support or enhance the functionality of the current development.

Stage 3: Implementation

In this phase, business models are formed around technology packages that derive value from various IP assets, including patents, utility models, industrial designs, layout designs of integrated circuits, trade secrets, trademarks, and copyrights. Formal agreements play a crucial role in maintaining confidentiality and defining rights. IP management here is focused on commercializing the innovation, tracking its use, and managing the licensing or deployment of protected intellectual property.

Institutional arrangements for IPR management vary depending on the context, but generally involve a combination of elements to effectively handle the complex process of protecting and leveraging intellectual property. Here's a breakdown of key components:

1. Dedicated Office

Technology Transfer Office (TTO): This is the most common setup, often found in universities and research institutions. A TTO acts as a central hub for managing IPR, including invention disclosure, patent filing, licensing, and commercialization strategies.

Innovation Centres: Some institutions may have dedicated innovation centres that focus on broader aspects of innovation beyond just IPR, but still incorporate IP management as a key aspect.

2. Personnel and Expertise

IP Specialists: TTOs or Innovation Centers typically employ IP specialists, including patent attorneys, licensing experts, and technology transfer professionals. These individuals handle the legal, technical, and financial aspects of IPR management.

Researchers and Faculty: Researchers and faculty members play a crucial role in identifying and disclosing inventions, contributing to the IP management process.

External Collaborators: Lawyers, consultants, and industry partners can be brought in for specific expertise and support as needed.

3. Policies and Procedures

IPR Policy: A formal policy outlining the institution's approach to IPR, including ownership rights, revenue sharing, and technology transfer guidelines.

Disclosure and Invention Management Procedures: Clear procedures for researchers to disclose potential inventions to the TTO or Innovation Center, ensuring timely identification and evaluation of IP assets.

Commercialization Strategies: Establishing strategies for bringing innovations to market, such as through licensing, joint ventures, or spin-off companies.

4. Collaboration and Communication

Internal Collaboration: Effective communication and collaboration between researchers, IP specialists, and other stakeholders within the institution is crucial for successful IPR management.

External Partnerships: Building partnerships with universities, research institutions, private companies, and government agencies can expand resources, expertise, and commercialization opportunities.

5. Resources and Funding

Dedicated Funding: Allocating sufficient resources to support the TTO or Innovation Centre's operations, including staff salaries, legal fees, and patent filing costs.

Incentives for Researchers: Mechanisms to incentivize researchers to disclose inventions and actively participate in the IPR management process, such as revenue-sharing models.

Effectiveness of Institutional Arrangements

The effectiveness of institutional arrangements for IPR management depends on several factors, including:

- Competence and expertise of the personnel involved.
- Clarity and comprehensiveness of the IPR policy and procedures. Adequate resources and funding for the TTO or Innovation Centre.
- Strong communication and collaboration within the institution and with external partners. By implementing these key elements and adapting them to their specifi c context, institutions can create eff ective arrangements for IPR management, enabling them to protect and leverage their innovations for broader benefi t.

IP Management Process

These six steps briefly outline the suggested flow of an IP management process:

Step 1: Identify the person(s) or form an office that will be responsible for handling IP issues within the organization: Although there are several functional units within most state university, it is recommended that a centralized contact be identified from which IP information can be managed and brokered.

Step 2: Establish a disclosure process: The entry point for identifying and documenting IP is the disclosure form. It is recommended that these forms be submitted and revised during periodic reviews of projects or other work activities being conducted by state university employees or by contractors working for or on behalf of the state.

Step 3: Screen and review: The screening and review process should include at minimum a review of prior art (when applicable); assessment of the field of application; identification of the key features and benefits of the potential IP; identification of the expected users and key stakeholders; and assessment of the IP's value to the state university. Technology assessment (review of competitors, market analysis, and review of similar and substitute technology) should be conducted on a case-by-case basis.

Step 4: Make the decision: It is important that the decision-makers consider the various opportunities presented in terms of social impact, financial impact, and overall long- termimpact on the university. For the state university, the decision will also be affected by the sources of project or activity funding and their requirements regarding work used on projects or other activities performed by state university employees or by contractors working for or on behalf of the state university **.**

Step 5: Performing technology transfer. It is important to determine how the technology will be transferred to the public. Details include:

1. Should the IP title and technology transfer responsibilities be taken by the contractor, the employee, or the state university
2. Should the IP be distributed freely and shared in the public domain?
3. Should the state university take title to the IP with a requirement to actively pursue licensing to third parties to transfer the technology?

Step 6: Monitoring. Depending on the technology transfer decision, it may be important to monitor and audit any agreements as well as any licensing compliance issues. Additionally, it will be important to measure the success of the IP management effort. These assessments may provide a trigger to initiate march-in rights (essentially, the right to acquire title if the IP holder is not satisfying certain criteria for commercialization), if applicable.

Mechanisms of IPR Management – Invention disclosure

Invention disclosure, the first step in the IP management process, is the crucial moment when an inventor shares their creation with the relevant authorities within their institution, usually a Technology Transfer Office (TTO) or Innovation Centre. This disclosure sets the wheels in motion for potential protection and commercialization of the invention.

Elements of a Strong Invention Disclosure

Descriptive Title: Clearly and concisely describe the invention in a way that captures its essence.

Technical Description: Provide a detailed explanation of the invention, its components, workings, and advantages over existing solutions. Drawings, diagrams, and schematics can be helpful supplements.

Background of the Invention: Briefly explain the problem or need addressed by the invention and its context within the relevant field.

Inventorship: Identify all individual contributors to the invention and their specific roles. Potential Applications and Market: Describe potential applications and the target market for the invention, demonstrating its commercial viability.

Supporting Documents: Include any relevant data, research results, prototypes, or other materials that support the invention's feasibility and novelty.

Here's how the invention disclosure process typically works:

1. Identification of Invention

The inventor identifies a novel idea, improvement, or solution to a problem that may have potential commercial or practical value.

This can happen through research, experimentation, or even serendipitous discovery.

2. Initial Documentation and Analysis

The inventor prepares a concise document describing the invention, including its technical details, potential applications, and competitive landscape.

This document may also include sketches, diagrams, or prototypes to further illustrate the invention.

Some institutions may offer internal resources or workshops to help inventors with this initial documentation.

3. Submission and Review

The inventor submits the disclosure document to the designated office within the institution, usually the Technology Transfer Office (TTO) or Innovation Centre.

The TTO or Innovation Centre reviews the disclosure to assess its patentability, commercial potential, and potential conflicts with existing intellectual property.

This review may involve legal counsel, patent attorneys, or technical experts depending on the complexity of the invention.

4. Decision and Next Steps

Based on the review, the TTO or Innovation Centre makes a decision on how to proceed.

This may involve pursuing patent protection, seeking funding for further development, negotiating licensing agreements with companies, or disclosing the invention publicly without seeking protection.

The inventor is informed of the decision and may be involved in discussions about the next steps.

Benefits of Effective Invention Disclosure

Early Evaluation: Promptly brings the invention to the attention of IP specialists who can assess its patentability and commercial potential.

Prior Art Search: Facilitates a search for existing patents or published material that might impact the invention's novelty and patentability.

Filing Strategy Development: Based on the invention disclosure, experts can determine the best course of action, such as patent filing, licensing, or further research and development.

Documentation and Record Keeping: Creates a formal record of the invention, ownership, and development timeline, essential for legal purposes and potential disputes.

Challenges in Invention Disclosure

Fear of Disclosure: Inventors may be apprehensive about sharing their ideas due to concerns about plagiarism, loss of ownership, or competition.

Lack of Awareness: Researchers may not understand the importance of invention disclosure or the proper procedures to follow.

Complexities of IP Law: Navigating the legal nuances of patentability and ownership can be intimidating for non-experts.

Best Practices for Encouraging Disclosure

Education and Awareness: Regularly educate researchers on the importance of disclosure and the benefits of IP protection.

Simplified Procedures: Develop user-friendly disclosure forms and online platforms to make the process easy and accessible.

Confidentiality Assurances: Implement robust confidentiality measures to protect inventors' ideas during the disclosure and evaluation process.

Incentives and Rewards: Offer recognition, financial rewards, or other incentives to motivate researchers to disclose inventions.

IP Portfolio Management

An intellectual property (IP) portfolio is a collection of an organization's intangible assets, including patents, trademarks, copyrights, trade secrets, and geographical indications. Effective IP portfolio management is crucial to maximize the value of these assets and drive business success. Here's a breakdown of key elements involved:

1. Inventory and Assessment

Cataloguing and documenting: Identifying all IP assets within the portfolio, including details like filing dates, scope of protection, and current status.

Evaluating potential: Assessing the commercial viability, competitive advantage, and potential revenue generation of each asset.

Categorization and prioritization: Grouping similar assets for efficient management and prioritizing high-value assets for focused investment.

2. Protection and Maintenance

Maintaining compliance: Ensuring timely renewal of patents, trademarks, and copyrights to avoid expiration and loss of protection.

Monitoring infringement: Actively monitoring the market for potential infringements and taking necessary legal action to protect your IP rights.

Maintaining records and evidence: Keeping detailed records of invention disclosures, development processes, and licensing agreements to support future claims.

3. Exploitation and Commercialization

Identifying opportunities: Exploring various options for commercializing your IP assets, such as licensing, joint ventures, spin-off companies, or internal product development.

Negotiating agreements: Engaging in negotiations with potential partners to secure favourable terms for licensing or other commercialization deals.

Developing marketing strategies: Creating effective marketing campaigns to raise awareness and drive demand for products or services based on your IP assets.

4. Risk Management and Portfolio Optimization

Identifying potential risks: Assessing potential legal challenges, market changes, or technological advancements that could impact the value of your IP assets.

Diversification: Diversifying your portfolio across different types of IP and industries to mitigate risks and spread potential revenue streams.

Regular review and adaptation: Regularly reviewing your portfolio performance and adapting your management strategies based on market trends and internal needs.

Tools and Resources

IP management software: Specialized software can help you track, organize, and analyze your IP portfolio data, providing valuable insights for decision-making.

IP consultants and attorneys: Consulting with legal and technical experts can offer valuable guidance on various aspects of IP management, including protection, enforcement, and commercialization.

Industry associations and networks: Participating in industry associations and networks can provide access to resources, best practices, and potential partners for IP collaboration.

By implementing effective IP portfolio management strategies, organizations can:

Increase revenue and profitability: Generate income through licensing, product sales, and other forms of IP exploitation.

Enhance brand reputation and competitive advantage: Build a strong brand identity and differentiate themselves from competitors through unique IP assets.

Boost innovation and creativity: Encourage further innovation by recognizing and rewarding the value of intellectual property within the organization.

Protection and Management of Biological Resources

National Biodiversity Act (2002)

The National Biodiversity Act (2002) is a landmark legislation enacted by the Indian government to conserve and sustainably utilize the country's vast biodiversity. This comprehensive act aims **to:**

Protect India's rich biological diversity: Recognizing the immense value of the country's diverse ecosystems and species, the act provides a framework for their conservation and sustainable use.

Ensure equitable sharing of benefits: The act emphasizes fair and equitable sharing of benefits arising from the use of biological resources and associated knowledge, addressing historical imbalances and recognizing the rights of local communities.

Promote sustainable use: The act encourages the utilization of biological resources in a way that ensures their long-term conservation and benefits future generations.

Biodiversity and Biological Resource

Biodiversity: Biodiversity has been defined under Section 2(b) of the Act as "the variability among living organisms from all sources and the ecological complexes of which they are part, and includes diversity within species or between species and of eco-systems".

Biological Resource: The Act also defines, Biological resources as "plants, animals and micro- organisms or parts thereof, their genetic material and by-products (excluding value added products) with actual or potential use or value, but does not include human genetic material."

Key Features of the Act

Establishment of Biodiversity Authority: The act establishes three levels of biodiversity authority: National Biodiversity Authority (NBA), State Biodiversity Boards (SBBs), and Local Biodiversity Management Committees (LBMCs). These authorities play crucial roles in implementing the act's provisions, including granting approvals for bioresource access and benefit-sharing negotiations.

Access and Benefit Sharing (ABS): The act regulates access to India's biological resources and associated knowledge by foreign individuals and companies. Prior informed consent of local communities and equitable sharing of benefits derived from research or commercial utilization are mandatory requirements.

Protection of Traditional Knowledge: The act recognizes and protects the traditional knowledge of local communities associated with biological resources. Communities have the right to decide the terms of access and benefit sharing for their knowledge.

Biodiversity Management Committees: The act empowers local communities through the establishment of Local Biodiversity Management Committees (LBMCs). These committees manage and conserve biodiversity at the local level, including developing Biodiversity Management Plans.

Financial Resources: The act establishes the National Biodiversity Fund to support biodiversity conservation activities, including research, capacity building, and awareness programs.

The act envisaged a three-tier structure to regulate the access to biological resources:

- The National Biodiversity Authority (NBA)
- The State Biodiversity Boards (SBBs)
- The Biodiversity Management Committees (BMCs) (at local level)

The Act allocates dedicated funds and a separate budget to the designated authorities for conducting research related to the country's biological and natural resources. These authorities are responsible for monitoring the use of biological resources, ensuring their sustainable utilization, overseeing financial investments related to these resources, and managing the returns appropriately.

They are also empowered to identify and notify endangered species, and to either prohibit or regulate their collection. Additionally, they can assign specific institutions to serve as repositories for various types of biological materials.

Under the Act, all offences are considered cognizable and non-bailable.

Exemptions under the Act include

- Indian biological resources that are commonly traded as commodities are exempt, provided they are used solely for commercial trade and not for any other purpose.
- Traditional uses of Indian biological resources and related knowledge are excluded, as well as their use in collaborative research projects involving Indian and foreign institutions, given such projects have the approval of the central government.
- Exemptions also apply to users like farmers, livestock rearers, beekeepers, and traditional practitioners such as vaids and hakims, for purposes related to their customary practices.

Protection of Plant Varieties and Farmers' Rights Act (PPVFRA)

The Protection of Plant Varieties and Farmers' Rights Act (PPVFRA) 2001 is a landmark legislation in India, balancing the protection of intellectual property for plant breeders with the recognition and safeguarding of the rights of farmers. It aims to:

Encourage development of new plant varieties: By providing a system for protecting the intellectual property rights of plant breeders through registration and granting exclusive rights, the act incentives research and development in the agricultural sector.

Facilitate the growth of seed industry in the country which will ensure the availability of high quality seeds and planting material to the farmers.

Objectives of the PPV & FR Act, 2001

To create a robust legal framework that safeguards plant varieties and secures the rights of both farmers and plant breeders, while promoting the creation of new plant varieties.

1. To acknowledge and uphold farmers' rights for their role in preserving, enhancing, and providing plant genetic resources that contribute to the development of new varieties.
2. To boost agricultural progress by protecting breeders' rights and encouraging investment in plant variety research and development across both public and private sectors.

Protect farmers' rights: Recognizing the crucial role of farmers in conserving and improving plant genetic resources, the act grants farmers specific rights, including:

Right to save, use, exchange, sell, or sow their farm-saved seeds: This ensures farmers remain self-sufficient and have the freedom to utilize traditional seed saving practices.

Right to register indigenous and traditionally evolved varieties: Farmers can register unique varieties they have conserved or developed, securing recognition and legal protection for their contributions.

Right to fair and equitable compensation: If a registered variety developed by a breeder infringes on a farmer's existing variety, the act provides mechanisms for seeking compensation.

Key Features of the Act

Registration System: Breeders can register new plant varieties with the PPV&FR Authority following specific criteria for novelty, distinctiveness, uniformity, and stability.

Exclusive Rights: Registered breeders have exclusive rights to produce, sell, import, and export the registered variety for a specific period (15 years for most crops).

Farmers' Rights Register: A separate register exists for the registration of indigenous and traditionally evolved varieties by farmers.

Dispute Settlement Mechanism: A specific mechanism is established to resolve disputes related to the infringement of farmers' rights or breeders' rights.

National Fund for Plant Genetic Resources: The act establishes a fund to support the conservation and utilization of plant genetic resources, including funding research and development activities.

General functions of the Authority

1. Registering new plant varieties, including essentially derived varieties (EDVs) and existing (extant) varieties.
2. Formulating DUS (Distinctiveness, Uniformity, and Stability) testing protocols for newly introduced plant species.
3. Developing methods for the identification, characterization, and record-keeping of registered plant varieties.
4. Providing mandatory cataloguing services for all categories of plant varieties.
5. Recording, classifying, and indexing varieties developed and conserved by farmers.
6. Acknowledging and honoring individual farmers or farming communities—especially from tribal and rural backgrounds—who contribute to the conservation, enhancement, and safeguarding of valuable plant genetic resources and their wild relatives.
7. Keeping and updating the National Register of Plant Varieties.
8. Managing and preserving plant genetic material in the National Gene Bank.

Rights Under the Act

Breeder's rights: Breeders have exclusive authority over the production, sale, marketing, distribution, import, and export of the protected plant variety. They are allowed to appoint agents or licensees to carry out these activities and can seek civil remedies if their rights are violated.

Researcher's Rights: Under the Act, researchers are permitted to use any registered plant variety for experiments or research activities. They can also use it as a base for developing a new variety. However, if the original variety is used repeatedly in the development process, the breeder's prior consent is necessary.

Farmers' Rights

1. A farmer who has developed a new variety is eligible to register and receive protection just like any breeder.
2. Farmers' varieties can be registered as extant varieties.
3. Farmers are permitted to save, reuse, sow, resow, exchange, and share seeds or farm produce of a protected variety as they did before the Act came into effect. However, they cannot sell such seeds under a brand name.
4. Farmers who conserve plant genetic resources, such as traditional varieties or wild relatives of cultivated crops, are eligible for recognition and rewards.
5. If a registered variety fails to perform as claimed, farmers are entitled to compensation under Section 39(2) of the Act.
6. Farmers are not required to pay any fee for proceedings under the Act, whether in front of the Authority, Registrar, Tribunal, or High Court.

Different Rights of Farmers: Farmers' Rights are essential for sustaining crop genetic diversity, which underpins food and agricultural production worldwide. These rights acknowledge and support farmers' traditional role in conserving and enhancing crop varieties, and reward their contribution to the global pool of genetic resources.

1. **Right to Seed**: Farmers retain the right to save, exchange, and use seed, similar to traditional practices before the Act. However, they cannot sell the seed in branded packaging under the name of the protected variety.
2. **Right to Register Varieties**: Farmers can register their own varieties, just like commercial breeders. These varieties must meet criteria of

distinctiveness, uniformity, and stability (DUS), but not necessarily novelty. This provision uniquely allows farmers to claim plant breeder rights for their traditionally cultivated or evolved varieties.

3. **Right to Reward and Recognition**: The Act establishes a National Gene Fund to acknowledge and reward farmers who have contributed to plant variety development or conservation. Contributions to this fund come from benefit-sharing fees paid by breeders. The exact implementation mechanism is left to the Authority created under the Act.
4. **Right to Benefit Sharing**: The National Gene Fund also facilitates benefit sharing. When a variety is registered, the Authority invites claims for benefit sharing. Any individual or organization can submit claims, and rewards are given to farmers or communities who can prove their contribution to the development or conservation of the variety.
5. **Right to Information and Compensation for Crop Failure**: Section 39(2) mandates that breeders must provide performance expectations of their varieties. If a variety fails to meet the claims, farmers can apply for compensation, ensuring accountability from seed companies.
6. **Right to Compensation for Undisclosed Use of Traditional Varieties**: If a breeder uses a community's traditional variety without acknowledging the source, compensation can be claimed through the Gene Fund. NGOs, individuals, or government institutions may submit claims on behalf of the affected community.
7. **Right to Adequate Availability of Registered Material**: Breeders are required to ensure public access to the registered variety at a fair price. If this is not done within three years, anyone can apply for a compulsory license, allowing third parties to produce and sell the variety.
8. **Right to Free Services**: Farmers are exempt from all fees related to variety registration, testing, renewal, opposition, and legal proceedings under the Act.
9. **Protection Against Unintentional Infringement**: In recognition of the low awareness levels among some farmers, the Act protects them from legal penalties if they unknowingly infringe on a breeder's rights and can prove a lack of knowledge.

Challenges and Opportunities

Despite its positive strides, the PPVFRA faces challenges:

Awareness and Capacity Building: Enhancing awareness among both farmers and breeders about the act's provisions and their rights is crucial.

Effective Implementation: Strengthening infrastructure and capacity of the PPV&FR Authority for efficient registration, dispute resolution, and enforcement is essential.

Balancing Interests: Reconciling the interests of breeders seeking commercial returns with the rights and livelihoods of farmers remains an ongoing challenge.

Despite these challenges, the PPVFRA offers a significant opportunity to achieve progress in

Indian Agriculture

- **Sustainable seed systems:** Empowering farmers with the freedom to save and exchange seeds can strengthen local seed systems and promote biodiversity.
- **Encouraging Innovation:** Granting rights to plant breeders promotes ongoing research and the development of improved crop varieties to meet diverse agricultural demands.
- **Equitable Benefit Sharing:** Acknowledging the role of farmers ensures they receive a fair share of the benefits generated from new agricultural technologies and plant varieties.
- **Safeguarding Traditional Knowledge:** The Act provides recognition and protection to the traditional knowledge held by local communities regarding biological resources. These communities have the authority to determine the conditions for access and benefit-sharing related to their knowledge.
- **Local Biodiversity Governance:** The Act strengthens the role of local communities by establishing Local Biodiversity Management Committees (LBMCs). These committees are responsible for conserving and managing biodiversity at the grassroots level, including preparing and implementing Biodiversity Management Plans.
- **Financial Resources:** The act establishes the National Biodiversity Fund to support biodiversity conservation activities, including research, capacity building, and awareness programs.

Agencies Responsible for Managing Biological Resources in India India employs a multi-layered framework for the governance and conservation of its rich biodiversity, with various institutions functioning at different levels. Here's an overview of the major stakeholders:

National Level

National Biodiversity Authority (NBA)

Serving as the top regulatory authority, the NBA operates under the Ministry of Environment, Forest and Climate Change (MoEFCC). It is tasked with enforcing the provisions of the Biological Diversity Act, 2002. The NBA plays a central role in regulating access to biological resources and ensuring fair benefit-sharing (ABS) related to traditional knowledge. It also issues guidelines and supervises the operations of State Biodiversity Boards (SBBs).

MoEFCC: Overall policy guidance and support for biodiversity conservation through various divisions like Wildlife, Forests, and Biodiversity & Climate Change.

National Gene Bank: Established by the Indian Council of Agricultural Research (ICAR), it safeguards genetic diversity of agricultural crops and important plant species.

State Level

State Biodiversity Boards (SBBs): These boards operate under the NBA's guidance and implement the Act at the state level. They regulate ABS within their states, approve bioresource access applications, and manage Biodiversity Management Committees (BMCs).

Forest Departments: Play a crucial role in managing protected areas, wildlife conservation, and biodiversity within forest ecosystems.

State Agricultural Universities and Research Institutes: Conduct research on plant genetic resources, develop new varieties, and contribute to conservation efforts.

Local Level

Biodiversity Management Committees (BMCs): Established by local bodies at the village level, these committees are responsible for managing biodiversity within their jurisdiction. They prepare Biodiversity Management Plans, conserve local resources, and promote sustainable practices.

Panchayati Raj Institutions: Village panchayats and other local governance bodies play a vital role in involving communities in biodiversity management and ensuring their rights are protected.

Additional Players

Research Institutions and Universities: Conduct research on various aspects of biodiversity, contributing to knowledge generation and conservation strategies.

Non-Governmental Organizations (NGOs): Advocate for biodiversity conservation, work with communities, and raise awareness about environmental issues.

Indigenous Communities and Local People: Custodians of traditional knowledge and practices related to biodiversity, actively involved in conservation efforts.

Collaborations and Coordination

Effective management of biological resources requires active collaboration and coordination among these diverse agencies. The NBA plays a key role in facilitating communication, providing technical support, and ensuring consistency in implementation across states. Regular dialogue and joint initiatives between government agencies, research institutions, communities, and NGOs are crucial for achieving successful biodiversity conservation in India.

Access to Genetic Resources and Sharing of Benefits (ABS)

Access to Genetic Resources and Sharing of Benefits (ABS) is a complex and dynamic field within the broader framework of environmental protection and intellectual property rights. It revolves around two key principles:

1. **Access to Genetic Resources:** Countries with rich biodiversity, often developing nations, provide access to their genetic resources (plants, animals, microorganisms, their genes, and associated traditional knowledge) to researchers and companies from other countries.

2. **Sharing of Benefits:** In return for this access, researchers and companies share the benefits arising from the use of these resources, such as profits from commercial products, research findings, and technology transfer. This ensures fair and equitable compensation for the resource providers, often local communities and indigenous groups.

Why is ABS important?

Fairness: Ensures that countries and communities providing valuable genetic resources don't get exploited and receive fair compensation for their contributions.

Biodiversity Conservation: Incentivizes sustainable use of resources and promotes conservation efforts by providing benefits to local communities.

Innovation: Promotes research and development by facilitating access to genetic resources while ensuring responsible and equitable utilization.

Key Aspects of ABS

Prior Informed Consent (PIC): Resource providers must give their free and informed consent before any access is granted. This ensures transparency and safeguards local communities' rights.

Mutually Agreed Terms (MAT): The specific terms of access and benefit sharing are negotiated between the providers and users, ensuring fairness and addressing concerns of both parties.

Compliance with National Legislation: Each country has its own ABS framework, often aligned with the Nagoya Protocol, an international agreement on ABS.

Benefit Sharing Mechanisms: Benefits can be shared in various ways, such as monetary payments, technology transfer, capacity building, and research collaboration.

Challenges and Opportunities

Implementing effective ABS frameworks: Many countries lack the infrastructure and resources to effectively implement and enforce their ABS regulations.

Ensuring equitable benefit sharing: Guaranteeing fair distribution of benefits among communities and addressing concerns about biopiracy.

Balancing access and conservation: Striking a balance between facilitating research and development while protecting biodiversity and local knowledge.

Despite these Challenges, ABS offers Numerous Opportunities

Empowering Local Communities: Ensuring that local communities receive economic benefits and are actively involved in conserving and utilizing their biological resources.

Advancing Sustainable Development: Supporting innovation and research aimed at creating new products and technologies that benefit both biodiversity and the well-being of local populations.

Encouraging Knowledge Exchange and Collaboration: Fostering partnerships among researchers, businesses, and communities to promote shared learning and mutual gains.

With growing awareness and the continuous development of Access and Benefit-Sharing (ABS) frameworks, there is increasing potential to establish a fairer and more inclusive system for distributing the benefits derived from biodiversity.

The **Nagoya Protocol**, adopted under the Convention on Biological Diversity (CBD), serves as the primary international agreement governing ABS. It outlines the guidelines, procedures, and responsibilities for the just and equitable sharing of benefits from the utilization of genetic resources.

Protection, Management and Commercialisation of Grassroot and Farmers Innovations, Traditional and Indigenous Knowledge

Traditional and Indigenous Knowledge (TIK) and grassroots and farmers' innovations (GFIs)

Traditional and Indigenous Knowledge (TIK) and grassroots and farmers' innovations (GFIs) represent a vast repository of knowledge, practices, and solutions developed by communities over generations through lived experience and interaction with their environment.

Understanding their meaning, forms, and importance is crucial for sustainable development and equitable resource management.

TIK: Refers to the knowledge, innovations, and practices of indigenous and traditional communities, often passed down through generations orally or through cultural practices. It encompasses diverse areas like agriculture, medicine, resource management, and environmental conservation.

Indigenous Traditional Knowledge (ITK) refers to the localized knowledge systems that are specific to a particular culture or community. While terms like "knowledge" (objective facts), "belief" (spiritual or religious views), and "tradition" (customary practices) can be conceptually distinguished, they are often used interchangeably when describing Indigenous ways of knowing.

Indigenous knowledge is often viewed as a form of social capital, especially among economically marginalized communities. It serves as a critical resource, helping them sustain livelihoods, grow food, secure shelter, and maintain autonomy over their lives.

ITK is built over generations through lived experience and deep familiarity with the natural environment. However, this valuable knowledge is increasingly at risk of disappearing due to rising population pressures and the expansion of industrial activities.

It's important to note that while **traditional knowledge** is a broad term encompassing various types of long-standing community knowledge, **indigenous knowledge** specifically refers to the traditional knowledge systems of indigenous peoples.

Examples: Intercropping, Crop rotation etc.,

GFIs: Are locally developed solutions to address agricultural and environmental challenges faced by farmers and communities. They often involve adaptations to specific ecological conditions and utilize readily available resources.

Forms

TIK: Exists in various forms, including:

Agricultural practices: Crop selection, cultivation methods, pest management, traditional food processing techniques.

Medicinal knowledge: Use of plants and natural remedies for health and well-being.

Resource management: Traditional water harvesting, soil conservation practices, sustainable forestry methods.

Cultural practices: Ceremonies, rituals, and stories reflecting ecological knowledge and respect for nature.

Grassroots and Farmers' Innovations (GFIs) can appear in multiple formats, including:

Crop improvement: Development of new varieties, seed saving practices, organic pest control methods.

Water management: Innovative irrigation systems, water harvesting techniques, drought- resistant crops.

Land management: Soil fertility enhancement methods, agroforestry practices, erosion control techniques.

Livestock management: Traditional breeding practices, disease control methods, sustainable grazing techniques.

Importance of TIK and GFIs

TIK and GFIs provide solutions to global challenges: They propose valuable insights into sustainable agriculture, natural resource management, and climate change adaptation.

Contribute to food security and livelihoods: Knowledge of diverse crops, resilient varieties, and efficient farming practices can enhance food production and improve community livelihoods.

Promote biodiversity conservation: Traditional practices often emphasize respectful co- existence with nature, contributing to the conservation of ecosystems and species diversity.

Enhance resilience and adaptability: Local knowledge and innovations empower communities to cope with environmental changes and build resilient livelihoods.

Recognize and value local wisdom: Documenting and respecting TIK and GFIs empowers communities and promotes cultural diversity.

Importance of Traditional and Indigenous Knowledge

1. Gaining insight into Indigenous Traditional Knowledge (ITK) supports the continuation of farming practices, helping to prevent the loss of plant genetic diversity and degradation of the environment.
2. It plays a vital role in ensuring long-term food security and preserving the diversity and richness of animal species, plant varieties, and soil characteristics.
3. It links the survival of every human being to the wholeness of nature and its elements that support life.
4. It provides the concrete situations of communities in relation with the environment and provides practical solutions to the problems of the people.
5. Indigenous knowledge, along with western- based knowledge, helps create development solutions that are culturally acceptable to the society being helped.
6. Facilitate participatory development processes, foster socioeconomic resilience of local communities and enhance the comparative advantage of a developing country.

7. Indigenous science informs place-specific resource management and land-care practices important for environmental health of tribal and federal lands.
8. Can contribute to modern science and natural resource management.
9. It also improves conservation, restoration, and the sustainable use of nature, which benefits society at large.
10. Indigenous knowledge provides a crucial foundation for community-based adaptation and mitigation actions.
11. Indigenous knowledge systems helps in gathering, predicting, interpreting and decision-making in relation to climate variability and weather events.

Documenting, Registering, Protecting, and Commercializing Indigenous Knowledge (IK)

Documenting, registering, protecting, and commercializing indigenous knowledge (IK) presents a complex and nuanced set of challenges and opportunities. While these processes aim to reserve and value IK, several concerns and considerations essential to be addressed to ensure equitable benefits and prevent exploitation. Here's a breakdown of the key elements:

Systems of Documentation

The process of documenting traditional knowledge involves identifying, collecting, organizing, and recording TK in a structured way to help manage, preserve, utilize, share, and protect it, based on specific goals and needs.

The documentation process generally includes three main stages:

- **Pre-documentation Phase:** This stage involves thorough planning, evaluating available approaches, setting clear objectives, and engaging in consultations with indigenous and local communities, along with other relevant stakeholders.
- **Documentation Phase:** At this point, traditional knowledge is actively gathered and systematically organized, following the strategies and actions outlined during the planning stage.
- **Post-documentation Phase:** This stage includes managing the documented data or system, such as a database or register. It involves overseeing how the recorded TK is accessed and used, ensuring proper governance of the information collected.

Key Objectives of TK Documentation

The process is intentional and strategic, serving several purposes such as:

- Establishing legal ownership or recognition of traditional knowledge (positive rights).
- Preventing misappropriation or wrongful intellectual property claims over TK (defensive protection).
- Making TK more accessible and systematically available for a broader audience, including researchers, students, and entrepreneurs.
- Facilitating the creation of new intellectual property through validation and collaborative research.
- Safeguarding and promoting traditional knowledge to ensure its transmission to future generations.
- Supporting the development and implementation of benefit-sharing mechanisms.
- Using TK to achieve specific community goals, such as education, cultural preservation, and raising awareness.

Method of Data Collection: Traditional knowledge is primarily gathered directly from communities (in situ) through interviews, discussions, observations, photography, audio- visual recordings, and other forms of engagement. methods for collection of ITK. It depends on type of ITK, situation, people, social system, cultural values and other aspects.

1. **Documentation of oral histories**: To preserve the indigenous practices in the form of pictures, audio recordings, video recording and notes etc.
2. **Agro-ecosystem analysis**: It is a thorough analysis of an agricultural environment which considers aspects from ecology, sociology, economics, and politics with equal weight.
   a) **Mapping (historical, social etc)**: Mapping is the creation of maps, a graphic symbolic representation of the significant features of a part of the surface of the earth.
   b) **Transect walk:** A transect walk is a tool for describing and showing the location and distribution of resources, features, landscape, main land uses along a given transect.
3. **Manual discriminative analysis**: In this process extension worker asks farmers to discriminate practices and find rationality.

4. **Decision tree analysis**: It is a decision support tool that uses a tree-like graph or model of decisions and their possible consequences, including chance event outcomes, resource costs, and utility.
5. **Use of local resource persons:** To get the information from the local people of the community.
6. **Analysis of journals and newspapers**: Analysis of journals and newspapers to know about the indigenous practices of the local community can be done.
7. **Continuous interactions during on-farm experiments**: Experts build a rapport with the local people during on farm experiments so that they can know about their way of living and their local practices.
8. **In-depth interview of farmers:** Farmers are interviewed about their local resources and their use for daily livelihood and problems faced.
9. **SWOT analysis**: It is a structured planning method used to evaluate the strengths, weaknesses, opportunities and threats involved in practices used by the indigenous community.

Types of Documentation

1. Documenting large variety of practices without scientific validation.
2. Documenting prevalent practices and comparing them with traditional ones.
3. Documenting the practices evolved to mitigate specific problems of farming or for sheer survival under conditions of ecological and economic stress.

Methods and Techniques of Documenting ITK

1. Notes; 2. Photos; 3. Audio-recordings; 4. Video-recordings

The Indigenous Knowledge (IK) Registration System

The Indigenous Knowledge (IK) Registration System formerly known as the National Recordal System (NRS) is an initiative of the Department of Science and Innovation (DSI) and is implemented in close collaboration with the CSIR, Software Architecture and Solutions Research Group under the Next Generation Enterprises and Institution (NGEI) Cluster. The aim of the IK Registration System is to facilitate the capturing, storage, curation, protection, promotion, and the management of Indigenous Knowledge (IK) within the context of the Protection, Promotion, Development and Management of Indigenous Knowledge Act 6 of 2019.

Systems: Some countries have established registers for IK, either standalone or integrated with intellectual property (IP) systems.

Challenges: Risk of rigid categorization, potential appropriation by others, and difficulty in defining ownership and boundaries of IK.

Benefits: Can provide communities with legal recognition and a basis for claims against misuse, and support commercialization efforts.

Systems of Traditional Knowledge Protection

Reasons to Protect Traditional Knowledge

1. Equity,
2. Biodiversity conservation,
3. Preservation of traditional practices,
4. Prevention of biopiracy, and.
5. TK's significance in development traditional knowledge. They are:

Positive protection involves granting traditional knowledge (TK) holders the legal authority to act against any unauthorized or inappropriate use of their knowledge. A framework for such protection should ensure the following:

1. Acknowledgement of the significance of traditional knowledge and the encouragement of respect for such systems.
2. Addressing the actual concerns and needs of traditional knowledge holders.
3. Prevention of unauthorized usage and unfair exploitation of traditional knowledge.
4. Safeguarding and encouraging innovation and creativity rooted in tradition.
5. Strengthening traditional knowledge systems and empowering their custodians.
6. Ensuring fair sharing of benefits derived from the utilization of traditional knowledge.

7. Encouraging the use of traditional knowledge in grassroots development efforts.

Defensive protection focuses on stopping unauthorized individuals or entities from wrongfully obtaining intellectual property rights over traditional knowledge. For such protection to be effective, it should:

1. Apply clear standards of prior art to traditional knowledge to prevent false novelty claims.
2. Establish systems that ensure traditional knowledge classified as prior art is documented and accessible to patent examiners and other relevant authorities.

Both positive and defensive protection methods are considered essential and should be used together, as no single approach is sufficient to fully safeguard traditional knowledge.

Legal principles Relevant to Traditional Knowledge Protection Include

1. **Prior Informed Consent (PIC):** Traditional knowledge holders must be properly informed and consulted before others use their knowledge.
2. **Equitable Benefit Sharing:** This ensures a fair distribution of benefits between traditional knowledge holders and users, balancing public and private interests.
3. **Unfair Competition:** Actions that mislead the public—such as falsely claiming a product is traditionally made or endorsed—can be challenged under unfair competition laws.
4. **Patents:** Innovations developed within traditional systems can be protected through the patent system.
5. **Distinctive Signs:** Trademarks, certification marks, collective marks, and geographical indications can protect traditional symbols, terms, and signs that are linked to traditional knowledge.
6. **Customary Laws:** Indigenous laws and traditional practices guide how communities create, use, and pass on their knowledge.

Non-IPR Approaches to Protecting Traditional Knowledge Include

1. **Environmental Protection:** The 1994 UN Convention to Combat Desertification recognizes the value of traditional knowledge in managing ecosystems and calls for benefit-sharing from its commercial use.

2. **Healthcare:** The World Health Organization, in its Alma Ata Declaration on Primary Health Care, acknowledged the importance of traditional knowledge in healthcare systems.
3. **Trade and Development:** The 2001 Doha Declaration by the World Trade Organization (WTO) instructed the TRIPS Council to consider and address issues surrounding the protection of traditional knowledge.
4. **Food and Agriculture:** The International Treaty on Plant Genetic Resources for Food and Agriculture highlights the importance of acknowledging farmers' rights and conserving traditional knowledge linked to plant genetic resources used in agriculture and food production.

Options: Sui generis systems specific to IK are being developed, alongside exploring potential adaptation of existing IP frameworks.

Challenges: Difficulty in proving novelty and distinctiveness of IK, legal complexities of communal ownership, and potential conflict with existing traditional practices.

Importance: Safeguards IK from unauthorized use and promotes fair compensation for communities.

Commercialization of Traditional Knowledge

What can be Commercialized?

1. Practices
2. Knowledge (e.g. may in the form of books)
3. Treatment practices
4. Products
   - Prepared by Practitioners
   - Prepared by Industry

Why Commercialization?

1. To make Traditional Knowledge & its benefits available to large section of the society
   1. To convert "tacit" knowledge to "explicit" knowledge
   2. To be a global player

Commercialization: Commercialization to be effective we need to understand

- Consumer angle
- Needs and demands
- Steps involved in product development
- How to compete globally: Wedding of new technologies with traditional systems/processes

Models: Community-controlled benefit-sharing agreements, licensing partnerships, and co- operative ventures offer promising avenues.

Challenges: Ensuring equitable distribution of benefits within communities, navigating complex legal contracts, and building capacity for marketing and business development.

Benefits: Generates income for communities, promotes sustainable utilization of IK, and incentivizes further innovation and knowledge sharing.

Key Considerations

Prior informed consent: Before documenting, registering, or commercially using indigenous knowledge (IK), it is essential to provide communities with complete information and obtain their explicit approval.

Community ownership and control: Mechanisms should ensure communities retain control over their IK and determine how it is used and shared.

Equitable benefit sharing: Any benefits arising from IK use must be shared fairly and transparently with the communities that hold this knowledge.

Respect for cultural values: Documentation and commercialization efforts should respect the cultural context and significance of IK.

Traditional Knowledge Digital Library (TKDL)

The TKDL is an innovative and first-of-its-kind initiative launched by the Government of India to safeguard and promote the country's extensive traditional knowledge. Launched in 2001, it is a joint effort by the Council of Scientific and Industrial Research (CSIR) in collaboration with the Ministry of AYUSH, Ministry of Science and Technology, and Ministry of Health.

This Digital Repository Plays a Vital Role in

- Safeguarding India's traditional medicinal systems, including Ayurveda, Unani, Siddha, Sowa Rigpa, and Yoga.

- Combating biopiracy by offering patent examiners worldwide access to documented prior art from Indian classical texts.
- Making traditional knowledge accessible globally in a structured, searchable, and standardized format.
    - The TKDL serves as both a protective mechanism against unauthorized patents and a valuable reference for researchers, policymakers, and intellectual property authorities.
    - The TKDL initiative preserves the country's traditional knowledge through digitalisation and aims to prevent the misappropriation of India's traditional medicine knowledge through patenting worldwide.
    - The TKDL database comprises knowledge from over 270 texts of Ayurveda, Unani, Siddha, Yoga and more recently Sowa Rigpa.

Why was TKDL set up?

- It was established to prevent exploitation and to protect Indian traditional knowledge at patent offices worldwide.
- The TKDL was initiated following the efforts taken by India to successfully revoke the turmeric and basmati patents granted by United States Patent and Trademark Office (USPTO) and neem patent granted by European Patent Office (EPO).

Objectives of TKDL

- Seeks to prevent the granting of patents for products developed utilising traditional knowledge where there has been little, if any, inventive step.
- Intends to act as a bridge between information recorded in ancient Sanskrit and patent examiners (with its database containing information in a language and format understandable to patent examiners).
- Facilitates access to information not easily available to patent examiners, thereby minimising the possibility that patents could be granted for "inventions" involving only minor or insignificant modifications.

Key Features of TKDL

Vast Database: It currently contains information on over 4.54 lakh formulations and practices extracted from ancient texts.

Multilingual Accessibility: Available in five languages (English, Japanese, French, German, and Spanish) besides Hindi, making it accessible to a wider audience.

Structured Classification: Information is classified using the Traditional Knowledge Resource Classification (TKRC) system, enabling efficient searching and retrieval.

Patent Examiner Support: Regularly used by patent offices worldwide, including India, to check for prior art during patent applications.

Third-Party Observations: Facilitates submission of third-party observations and pre-grant oppositions based on TKDL information.

Benefits and Impact

Protection of Intellectual Property: Serves as a vital tool in preventing misappropriation of Indian TK and safeguarding the interests of traditional knowledge holders.

Knowledge Sharing and Promotion: Promotes wider access to and appreciation of India's rich traditional medical and wellness systems.

Scientific Validation: Encourages scientific research and validation of traditional knowledge, opening doors for potential innovation and commercialization.

Community Empowerment: Gives traditional knowledge holders a voice and recognition for their contribution to healthcare and well-being.

How TKDL Preserves Information?

- India's traditional medicinal knowledge exists in languages such as Sanskrit, Hindi, Arabic, Urdu, Tamil, etc, that too in ancient local dialects that are no longer in practice.
- Thus, the published traditional knowledge literature is neither accessible nor understood by patent examiners at international patent offices.
- The TKDL has successfully addressed challenges related to language and formatting by methodically translating and organizing the content of ancient Indian medicinal texts—covering Ayurveda, Siddha, Unani, Sowa Rigpa, and Yoga—into five major international languages: English, Japanese, French, German, and Spanish. This has been achieved using advanced information technology tools and an innovative system known as the Traditional Knowledge Resource Classification (TKRC).
- Under existing approvals, access to the TKDL database is granted to global patent offices that have signed non-disclosure agreements with the Council of Scientific and Industrial Research (CSIR). Currently, 14 such patent offices have secured access to the TKDL.

- The CSIR-TKDL Unit actively submits third-party observations and pre-grant oppositions to challenge patent applications that misuse India's traditional knowledge. To date, TKDL evidence has led to the withdrawal, amendment, or rejection of 245 patent applications, thus safeguarding India's traditional heritage.

Community Biodiversity Registers (CBRs)

The traditional knowledge of local people regarding local biological resources, their medicinal value or any other use can be enlisted to provide comprehensive information to all, known as community biodiversity registers or people biodiversity registers.

For example, fertilizer aided agriculture is suffering from an intensification of pest and disease outbreaks, with many pests having acquired resistance. Croplands are becoming unproductive due to mineral deficiency, erosion, water logging, salinity etc. Many medicinal treesand herbs are on the verge of extinction. Forests are being depleted resulting in shortage of fuel wood and the village economy is in deep trouble.

Under such circumstances, Community Biodiversity Register (CBR) acts an important instrument that can be used by local people to record the resources in terms of biodiversity, traditional knowledge and good practices. A CBR helps creating a sense of ownership and also helps to safeguard indigenous knowledge, local crops and livestock resources. It helps to strive a balance between development and conservation to eliminate the unwanted consequences leading to the loss of important genetic resources and diversity.

Community Biodiversity Registers (CBRs) are innovative tools used to document, manage, and protect the biodiversity and associated traditional knowledge within a specific community. They have emerged as powerful instruments for empowering communities and promoting sustainable resource management. Here's a breakdown of their key features and significance:

What are CBRs?

A participatory process where communities systematically document and record the biodiversity resources (plants, animals, microorganisms) found within their traditional territory.

Includes associated traditional knowledge (e.g., medicinal uses, conservation practices) passed down through generations.

Serves as a community-owned database that empowers them to manage their resources and negotiate with external actors.

Key Features

Community-led: Created and managed by the community, allowing them to retain full ownership and authority over the information.

Participatory approach: All members of the community, regardless of age or gender, are encouraged to contribute their knowledge and perspectives.

Focus on traditional knowledge: Recognizes and valorizes the wisdom and practices accumulated by communities over generations.

Biodiversity documentation: Covers a wide range of resources, including flora, fauna, and their ecological interactions.

Dynamic process: Continuously updated and adapted as new knowledge is acquired or circumstances change.

Benefits and Significance

Empowerment: Gives communities control over their resources and knowledge, increasing their bargaining power and decision-making autonomy.

Conservation: Promotes sustainable use and conservation of biodiversity by raising awareness and fostering community stewardship.

Protection of traditional knowledge: Safeguards against biopiracy and exploitation of local wisdom by external actors.

Benefit sharing: Provides a basis for negotiating fair and equitable benefits with companies or researchers seeking access to resources or knowledge.

Decision-making tool: Informs community development plans, resource management strategies, and conservation initiatives.

People's Biodiversity Registers (PBRs)

People's Biodiversity Registers (PBRs) are a cornerstone initiative under the National Biodiversity Act (2002) in India. The People's Biodiversity Registers program is sponsored by WWF India as part of its Biodiversity Conservation Prioritisation Project. The program's name was changed from "Community" to "People's" Biodiversity to reflect that not all local knowledge is communally-generated and shared freely among community members.11 The program also involves two institutions in Bangalore, the Centre for Ecological Sciences of the Indian Institute of Science, and the Foundation for Revitalisation of Local Health Traditions.

PBRs preparation is the new concept under the Biological Diversity Act, 2002. The concept was defined in the BD Rules 2004 and State specific Rule. So far about 5000 PBRs prepared in India. But few states like Kerala has been completed the documentation of PBRs for all the Biodiversity Management Committees.

Aims

1. Document Local Biodiversity and Traditional Knowledge: PBRs comprehensively compile information on the availability and uses of biological resources found within a specific geographical area. This includes details on plants, animals, microorganisms, and associated traditional knowledge held by local communities.
2. Empower Local Communities: PBRs are developed and maintained by local communities themselves, fostering a sense of ownership and control over their resources and knowledge. This empowers communities to make informed decisions about resource use and participate in biodiversity conservation efforts.
3. Facilitate Access and Benefit Sharing (ABS): PBRs provide a clear understanding of local biodiversity and knowledge, thereby aiding in the implementation of the ABS framework outlined in the National Biodiversity Act. This ensures fair and equitable sharing of benefits arising from the utilization of local resources and knowledge.
4. Promote Sustainable Resource Management: By documenting traditional practices and conservation methods, PBRs contribute to sustainable resource management. This knowledge can inform biodiversity conservation plans and policies, leading to a more balanced relationship between communities and their environment.

Key Features of PBRs

Participatory Process: Local communities directly contribute their knowledge and expertise to the register.

Comprehensive Coverage: Covers all biological resources and associated traditional knowledge within the area.

Multilingual Documentation: Information is often recorded in local languages and translated for wider accessibility.

Continuous Revision: PBRs are living documents that are regularly revised to incorporate new findings and adapt to evolving conditions.

Public Access: Registers are publicly accessible, promoting transparency and informed decision- making.

The Program's Purposes are

- To document and preserve traditional knowledge for the benefit of current and future generations within village communities.
- To support the revival and strengthening of local knowledge by:
  (a) acknowledging the diversity of traditional practices;
  (b) recognizing and rewarding exceptional knowledge, skills, techniques, and conservation methods;
  (c) validating and promoting reliable traditional knowledge and sustainable resource management systems; and
  (d) encouraging the exchange of knowledge between communities to build collective capacity.
- To raise awareness among conservationists about the urgency of protecting endangered resources and safeguarding local communities' rights to those resources.
- To prevent the exploitation of local biodiversity and traditional knowledge by external entities, such as through unauthorized patents on adapted products, processes, or biological materials.

Importance of PBR

1. To create a biodiversity management plan aimed at protecting the area's current biodiversity, with special focus on conserving rare, endangered, and threatened (RET) species of both flora and fauna.
2. To establish a baseline database that records existing biodiversity, enabling future assessment and comparison.
3. To serve as a legal document addressing intellectual property rights (IPR) and patent- related issues, thereby empowering local communities by making them aware of their rights and the importance of biodiversity conservation.
4. PBRs play a foundational role in connecting the traditional knowledge and genetic resources of local communities with the benefits derived from them, ensuring that local people receive their fair share.

5. PBRs can also contribute to the planning and development of both rural and urban regions.

Key Features of PBRs

1. To control access to the country's biological resources in a way that guarantees fair and equitable benefit-sharing for the use of these resources and the traditional knowledge associated with them.
2. To promote the conservation and sustainable utilization of biodiversity.
3. To honor and safeguard traditional knowledge possessed by local communities concerning biodiversity.
4. To ensure that local communities receive a just share of benefits, as they are the custodians of biodiversity and its associated knowledge.
5. To identify and develop areas rich in biodiversity by designating them as biodiversity heritage sites.
6. To protect and restore species that are under threat.
7. To include state government institutions in the broader implementation of the Biological Diversity Act through the formation of appropriate committees.

The PBRs will Helps in

1. Preparing the Biodiversity Management plants for the conservation of biodiversity Declaration of Biodiversity Heritage Sites.
2. Conservation of threatened flora, fauna and endemic species of the area falling within its territorial jurisdiction.
3. Preparing the community & Indigenous protocols for biodiversity conservation.
4. Increasing the livelihoods of the local community who are depending on the biodiversity.
5. Documentation of potential biodiversity for ABS mechanism.
6. Effective management, promotion and sustainable uses.
7. Protection of rights including IPR over Biological resources and associated knowledge and systems.
8. Restrict the flora and fauna which are threatened, endemic, rare, etc.
9. Restrict the activities which cause genetic erosion of affecting the ecosystem.

PBR covers 36 items - (Crop Plants, Fruit Plants, Fodder Crop, Weeds, Pests of Crops, Domestic Animals, Soil types, Medicinal Plants, Ornamental Plants, Timber Plant, Fisheries, Wild Biodiversity (Trees, Shrubs, herbs, Tubers, Grasses, Climbers), Aquatic Biodiversity, Wild Animals (Mammals, Birds, Reptiles, Amphibia, Insects, Others) etc. PBR register receives legal protection against misuse of and appropriation by outside agencies and individuals.

Registers (PBRs) based on Gadgil (1998)

People's Biodiversity Registers (PBRs) Encompass Ten Key Types of Information

1. Identification of various user groups that depend on local biological resources (referred to as the "Peoplescape").
2. Detailed mapping of the ecological habitats within the study area (the "Landscape").
3. A record of the ecological history of the area under study.
4. Documentation of the extent, spread, and nature of both individual and community knowledge regarding local plant and animal species, including their uses (the knowledge system).
5. Information on the population status, rarity, and distribution patterns of various living organisms.
6. Analysis of how biological resources are used economically, both for livelihood and commercial purposes.
7. Documentation of conservation and resource-use regulation efforts, initiated by either government bodies or local communities.
8. Insights into the developmental goals of the community and how these aspirations align with the local biodiversity.
9. Exploration of differences and common ground among local groups in managing natural resources.
10. Potential future strategies for managing natural resources in the area, with an emphasis on conserving biodiversity.

Important Issues Related to the PBR

1. It is to be undertaken in a participatory mode involving various sections of village society.

2. Information provided by people needs to be collated, analysed and cross-checked by the members of a Technical Support Group before documentation.
3. The PBR is an important base document in the legal arena as evidence of prior knowledge and hence careful documentation is necessary.
4. The document should be endorsed by the BMC and later publicized in the Janpad Panchayat/Gram Panchayat/Panchayat Samiti.
5. The document could be a very useful tool in the management and sustainable use of bio- resources.
6. The document can also be a very useful teaching tool for environmental studies at schools, colleges and university level.
7. The document should be periodically updated with additional and new information as and when generated.

Honeybee Network

The Honey Bee Network's central objective is ethical knowledge extraction. It maintains a grassroots knowledge database that collects and disseminates expertise from a wide range of individuals while observing ethical practices of credit, compensation, and accessibility.

The Honey Bee Network was founded by Anil Gupta in 1988 in India as a social justice initiative to fix the failures of top-down development initiatives led by the government. It aims to break down the ethical and professional dilemmas that prevent leaders from obtaining and disseminating knowledge from the country's poorest people. Changing the mode of how that knowledge is gathered and communicated forced a wholesale change on how policies meant to alleviate poverty and its problems are formulated.

1. The Honey Bee Network's central objective is ethical knowledge extraction.
2. It maintains a grassroots knowledge database that collects and disseminates expertise from a wide range of individuals while observing ethical practices of credit, compensation, and accessibility.
3. It strives to build a network that allows the benefits of its centralized database to be shared with knowledge contributors that may normally go ignored under the traditional practices of knowledge extraction.
4. Honey Bee Network, a multimedia database of local knowledge that is collected and circulated through grassroots outreach amongst knowledge-rich grassroots communities.

5. The organization is composed of "like-minded individuals, innovators, farmers, scholars, academicians, policy makers, entrepreneurs and non-governmental organizations" that facilitate the collection and dissemination of local knowledge.
6. The name is an intentional nod to honey bees that cross-pollinate by enriching the flowers they touch.

At the time, the prevailing assumption was to look to government and civil society to provide poverty alleviation solutions. Gupta instead observed through his original work that creativity and knowledge resided with the people living in the most severe conditions. People were innovative in their approach to the various problems they were facing. Local residents knew best how to deal with problems, whether they were living in flood prone areas, forests, or the dry deserts. The scarcity of resources did not stop people from figuring out solutions.

But there was an asymmetry in how the poor's knowledge was perceived, valued, or used, and sharing disadvantaged the poor. To tackle this issue, Gupta founded the Honey Bee Network— a multimedia repository of indigenous knowledge that is gathered and shared through grassroots efforts among communities rich in traditional wisdom. The organization is composed of "like-minded individuals, innovators, farmers, scholars, academicians, policy makers, entrepreneurs and non-governmental organizations" that facilitate the collection and dissemination of local knowledge. The name is an intentional nod to honey bees that cross- pollinate by enriching the flowers they touch.

Principles of Network

The Network is run based on three main guiding principles.

1. First, when new knowledge is collected, it must be shared back with the sources and the communities in their local languages. In the past, it was seen that academicians and researchers would obtain all the information from the grassroots level and distribute it in ways that were inaccessible to the knowledge producers.
2. The second principle is that the source of the knowledge must be acknowledged.
3. Finally, the knowledge holders must benefit from the success of their work and inventions in both fame and remuneration.

The principles we can learn from the life of honeybee are:

- Just as flowers don't object when bees collect their nectar or pollen, people should not object to their traditional knowledge being documented by others. However, it is essential that the original knowledge holders are properly credited, including their names and locations, and that their intellectual property rights are honored.
- Bees play a vital role in cross-pollination, enhancing biodiversity and sustaining nature's balance. Similarly, unless we communicate in local languages that people can easily understand, genuine connections between communities won't form, and peer-to- peer learning won't occur. Furthermore, whenever traditional knowledge leads to wealth creation whether through commercial or non-commercial means, with or without added value a fair portion of that benefit should be returned to the knowledge holders who made it possible.
- Before disseminating people's knowledge or bringing it in public domain, their prior informed consent should be taken.

Since its founding, the network's database of original inventions, presented through illustrations and with direct attribution to the original inventor, has grown to more than 100,000. The database is designed to be easy for local communities to use, secure, and in multiple local languages. The Network does grassroots outreach to local communities to share the solutions and also to encourage sharing of their ideas and knowledge, and publishes a newsletter in eight languages and distributed to 75 countries.

The Honey Bee Network has played a key role in securing patents for numerous grassroots innovations and in facilitating funding to help scale some of these inventions on a global level. In light of two major global trends—the shift toward more conservative governance with reduced focus on social development, and the push to decolonize design, innovation, and science—the Network is intensifying its efforts. It aims to expand its services by creating interactive platforms that enable local innovators to connect with peers, protect their intellectual property, attract interest in their products or services, and engage with potential investors.

Society for Research and Initiatives for Sustainable Technologies and Institutions (SRISTI)

SRISTI is a voluntary developmental organization dedicated to promoting and nurturing the creativity of grassroots communities and individual innovators. It focuses on supporting environmentally sustainable solutions to local challenges, which have been identified, developed, and disseminated through the Honey Bee Network over the past 33 years.

SRISTI upholds the five foundational principles of the Honey Bee Network:

1. Encouraging innovative practices in education by teachers, students, and other stakeholders at schools and colleges.
2. Fostering institutional innovations at the community and broader levels for effective resource management.
3. Enhancing the access of economically disadvantaged but knowledge -rich individuals to opportunities that enable self -sustaining development.
4. Promoting cultural creativity to nurture curiosity, cooperation, and compassion through art, literature, crafts, and other forms of expression.
5. Supporting technological and traditional knowledge -based innovations that address human, animal, plant, and environmental health, while also advocating for policy changes to promote affordable, sustainable innovations —especially focusing on youth, children , women, and the elderly.

The Honey Bee Network tries to

1. Provide a peer group of farmers, artisans, scientists, academics across language, culture and regional boundaries to nurture, critique and encourage innovative experimentation.
2. Break the nexus between the regions of high biodiversity and high poverty endowed with poor public as well as private infrastructure.
3. Provide access to information that can help improve productivity without increasing cost using other farmers' innovations. Most of the grassroots innovators do not have access to relevant information, which cripples their ability to raise resources and explore opportunities in different markets. Thus, poor demand for ecological and technological skills as well as their eco-friendly products forces them to become "unskilled labourer" in the urban houses and markets.
4. Galvanize existing institutions and community structures to inspire and sustain the curiosity and spirit of younger generation to pursue the path of experimentation and excellence in local eco-enterprises and natural resource management.

5. Resolve an ethical dilemma about sharing and protecting traditional as well as contemporary knowledge of individuals and communities evolved through conscious efforts but guided by different value systems without keeping people poor. The fact that we have not found many young healers indicates that younger generation does not find the career of healer or herbalist worth pursuing when it entails a life of penury though with a lot of goodwill in the community.
6. Generate a system for rewarding and providing incentives to the innovative individuals as well communities under the provision of several international and national agreements like Convention on Biological Diversity (Art 8(J)), International Convention to Combat Desertification (Art 16), etc.

Geographical Indications (GI) and Appellation of Origin

Geographical Indications (GIs) and Appellations of Origin (AOs) are intellectual property rights that protect the names and reputations of products originating from specific geographical locations, whose qualities and characteristics are essentially attributable to that place. Let's delve into their meaning, origin, and the Indian legal framework surrounding GIs.

Geographical Indication (GI)

A GI is a broad term that signifies any symbol or sign used on products that originate from a particular geographical location, where a specific quality, reputation, or characteristic of the product is inherently linked to that origin. It applies to goods that are agricultural, natural, or manufactured. For manufactured products, at least one stage—whether production, processing, or preparation—must occur in the defined geographical area.

1. It is clarified that a GI doesn't necessarily have to be a place name; for example, names like *Alphonso* and *Basmati* also qualify.
2. The term "goods" covers items from handicrafts and industrial sectors as well as food products.

Example: Basmati rice, Darjeeling Tea, Nagpur Oranges, Kolhapuri Chappal, Thirunelveli Halwa, Kanchipuram Sarees etc.

Appellation of Origin (AO): A stricter type of GI, often used for agricultural products (e.g., wines, cheese) where the specific qualities are primarily due to the unique environmental and cultural factors of the place. AOs have stricter regulations and often undergo a more rigorous registration process than regular GIs.

Appellations of origin are a special kind of geographical indication. The term is used in the Paris Convention and defined in the Lisbon Agreement. Article 2 of the Lisbon Agreement defines appellations of origin as"*(1)... the geographical denomination of a country, region, or locality, which serves to designate a product originating therein, the quality or characteristics of which are due exclusively or essentially to the geographical environment, including natural and human factors.*"

This definition suggests that appellations of origin consist of the name of the product's place of origin. However, it is interesting to note that a number of traditional indications that are not place names, but refer to a product in connection with a place, are protected as appellations of origin under the Lisbon Agreement (for example, Reblochon (cheese) and Vinho Verde (green wine)).

1. Both appellations of origin and geographical indications require a clear and specific connection between a product and its geographical origin, highlighting a quality or characteristic tied to that location. These labels help inform consumers about the product's origin and associated qualities.

2. The main distinction between the two lies in the strength of the connection— appellations of origin demand a more direct and substantial link to the place of origin.

3. For a product to be recognized under an appellation of origin, its qualities or characteristics must be entirely or primarily derived from its geographical location.

4. This typically implies that both the sourcing of raw materials and the processing of the product must occur within that specific geographic region.

5. In contrast, for geographical indications, even a single attribute—such as quality, characteristic, or simply reputation—connected to the geographic origin is enough for recognition.

6. Additionally, the raw material production or the processing of a GI-labeled product does not always need to take place fully within the designated area.

7. The term "appellation of origin" is frequently found in legal frameworks that provide dedicated or sui generis protection systems for geographical indications.

Origin

International Context

The concepts of Geographical Indications (GIs) and Appellations of Origin (AOs) received global recognition through the Agreement on Trade-Related Aspects of Intellectual Property Rights (TRIPS) under the World Trade Organization (WTO).

Geographical Indications of Goods (Registration and Protection) Act, 1999: The GI Act of 1999 is a unique legal framework passed by the Indian Parliament to safeguard geographical indications within the country. This legislation was introduced to align India's IP laws with the TRIPS Agreement, following its WTO membership. Darjeeling tea became the first product to receive GI status in India in 2004–05, and by August 2020, a total of 370 products had been registered.

Aims of the GI Act, 1999: The Act is designed to:

1. Prevent unauthorized use of geographical indications and protect consumers from misleading practices.
2. Establish a specific legal framework to address the interests of genuine producers of such goods.
3. Support and enhance the export potential of products bearing Indian geographical indications.

Key Provisions of the GI Act, 1999

1. Provides clear definitions for essential terms like "geographical indication," "goods," "producer," "registered proprietor," and "authorized user."
2. Mandates maintaining a GI register with two parts: Part A (registered GIs) and Part B (authorized user details).
3. Allows for GI registration in designated product categories.
4. Enables rejection of GI applications that do not meet required standards.
5. Empowers the Central Government to create rules for application procedures and evaluation criteria.
6. Requires all approved GI applications to be published and open to opposition.

7. Lists authorized users of registered GIs and outlines how they or the registered proprietors can take legal action against infringement.
8. Provides for stronger protection measures for specially notified products.
9. Forbids transferring or assigning GIs, as they are considered public property.
10. Bars the registration of GIs as trademarks.
11. Allows appeals against the Registrar's decisions to be made to the Intellectual Property Appellate Board under the Trademark law.
12. Specifies legal penalties for violations.
13. Details the benefits of registration and the rights granted.
14. Includes administrative requirements such as Registrar's powers, maintaining indexes, and addressing protection for homonymous GIs.

Benefits of GI Protection

Protection for Producers: Safeguards the reputation and value of products from specific regions, preventing misuse and unfair competition.

Consumer Protection: Ensures consumers get authentic products with specific characteristics and quality.

Economic Development: Promotes local economies and livelihoods of communities involved in GI production.

Cultural Heritage Preservation: Recognizes and protects the unique cultural and traditional practices associated with GI products.

Geographical Indication registration

Section 8 of the G.I. Act gives that a Geographical Indication may be registered regarding any or all of the goods, included in such types of goods as may be listed by the registrar. Moreover, regarding a particular area of a country, or a region or locality in that territory, as the case may be. According to the prescribed manner, the registrar may also classify the goods according to the international division of goods to register geographical indications and publish in an alphabetical index of various goods.

Exclusion

According to Section 9 of the Act, certain types of geographical indications (GIs) are not eligible for registration. These include:

1. Indications that may mislead or confuse the public.
2. Indications that violate any existing law.
3. Indications containing offensive, scandalous, or obscene material.
4. Indications that may offend the religious sentiments of any community or group in India.
5. Indications that, for other legal reasons, do not qualify for protection in a court of law.
6. Generic terms that have become common names.
7. Indications that falsely suggest the goods come from a different region, locality, or territory than they actually do.

Effects of Registration and Legal Protection

- Once a GI is registered, the owner and authorized users have the exclusive right to use that indication for the specified goods.
- Registration also enables them to file legal suits in case of infringement and claim damages.
- An infringement occurs when a GI is used on goods falsely suggesting they come from the registered origin or in a manner that results in unfair competition.
- If the GI is not registered, legal action can still be taken under the law of "passing off."
- A registered GI serves as initial proof of ownership and the validity of the claim in legal proceedings.
- The registration of a GI cannot be sold, assigned, licensed, or mortgaged, except when it is inherited after the death of an authorized user.
- Anyone who misuses a GI—by falsely applying it, tampering with the place of origin, or possessing equipment meant for falsification—may face legal consequences. This includes imprisonment for a minimum of six months (up to three years) and a fine ranging from INR 50,000 to INR 2,00,000.

- For repeat offenses, the punishment increases to a minimum of one year (up to three years) imprisonment and a fine of at least INR1,00,000 (up to INR 2,00,000).
- However, under special circumstances, a judge may reduce the sentence, but the reasons must be clearly stated in the judgment.
- Other punishable actions include falsely claiming a GI is registered, tampering with the official register, or misrepresenting any place as associated with the GI Registry.

Differences between GI and AO

Comparison Criterion	Geographical Indication	Appellation of Origin of Goods
Requirements for Designation	Any designation, including graphic or combined one, which makes it possible to identify goods as originating from the geographic location.	A verbal designation representing the name of the geographic location.
	A designation (not necessarily known), which makes it possible to identify goods as originating from the geographic location.	A designation, which became known as a result of its use in relation to goods.
Links of characteristics of goods with the geographic location	Quality, reputation, and other characteristics of goods, in relation to which a geographical indication may be registered, are largely related to its geographical origin.	Special properties of goods, in relation to which an appellation of origin of goods may be registered, are exclusively determined by the environmental conditions and (or) human factors peculiar to this geographical location.
Manufacture of products	At least one stage of manufacture of goods, which has material effect on formation of the characteristics of goods, should be performed in that geographical location.	All stages of manufacturing of goods, which have a material effect on formation of special properties of the goods, should be performed in that geographical location.
Protection mark	Special sign for a geographical indication	Special sign for an appellation of origin of goods

Genetically Modified Organisms (GMO), Agriculture and Biosafety

The Global Concerns on Use of Genetically Modified Organisms in Food and Agriculture Genetically modified organisms (GMOs) in food and agriculture continue to spark debate and raise concerns worldwide. While proponents highlight potential benefits like increased yields, pest resistance, and improved nutritional content, opponents express anxieties about potential health risks, environmental impact, and ethical considerations. Here's a closer look at some key concerns:

Health Concerns

Allergenicity: The introduction of foreign genes may create new allergens in GMOs, posing potential risks for allergic individuals.

Antibiotic resistance: The use of antibiotic-resistant genes in some GMOs may contribute to the spread of antibiotic resistance in bacteria, jeopardizing the effectiveness of these crucial medical tools.

Unforeseen effects: Long-term health impacts of consuming GMOs are still under investigation, raising concerns about potential unknown risks.

Environmental Concerns

Gene transfer: Unwanted transgene escape from GMOs to wild relatives could occur, potentially harming biodiversity and impacting ecosystems.

Pesticide use: Increased reliance on herbicide-resistant GMOs may lead to more herbicide use, harming wildlife and soil health.

Loss of biodiversity: Monoculture farming practices often associated with GMOs can contribute to a decline in biodiversity, impacting ecosystem resilience and natural pest control mechanisms. Ethical Concerns:

Corporate control: Large corporations heavily involved in GMO development and seed production raise concerns about farmer dependence and control over the food system.

Patenting life: Patenting of GMOs and seeds is seen by some as unethical, restricting access and potentially hindering food security in developing countries.

Insufficient transparency: There are concerns about the openness of GMO research and development processes, along with issues related to labelling and informing consumers about the presence of GMOs in food products.

Addressing Concerns

Rigorous scientific testing, robust regulatory frameworks, and open communication are crucial to address these concerns effectively. Ongoing research can shed light on potential risks and ensure GMOs are safe for consumption. Governments should implement transparent and strict regulations throughout the GMO development, testing, and marketing process. Open dialogue and information sharing with the public are essential to promote informed choices and address ethical concerns.

Moving Forward

The debate surrounding GMOs is complex and requires a nuanced approach. While acknowledging potential risks, it's important to consider the potential benefits GMOs could offer in addressing global food security challenges and increasing agricultural sustainability. Continued research, robust regulations, and open communication are key to ensuring the responsible development and use of GMOs, maximizing potential benefits while minimizing potential risks.

The National Seed Policy of India (2002)

The National Seed Policy of India, implemented in 2002, is a comprehensive framework aimed at enhancing the Indian seed sector's efficiency and promoting sustainable agricultural development. Here's a breakdown of its key objectives, thrust areas, and impact:

Objectives

Ensure availability of quality seeds: To guarantee sufficient supply of high-quality seeds of diverse varieties to Indian farmers.

Boost seed industry growth: To stimulate investment in research and development, production, and distribution of seeds to strengthen the Indian seed industry.

Protect farmers' interests: To safeguard the rights of farmers to save, use, exchange, and sell farm-saved seeds, while also promoting adoption of certified seeds.

Promote sustainable agriculture: To encourage adoption of improved varieties and best practices in seed production and management, contributing to sustainable agricultural practices.

Thrust Areas

Varietal Development and Plant Variety Protection: Focus on development of new and improved varieties, strengthening the plant variety protection system, and promoting private sector participation in research and development.

Seed Production: Improve infrastructure and production capacity for quality seeds, ensuring accessibility to farmers across the country.

Quality Assurance: Strengthen seed quality control mechanisms through certification systems and laboratory infrastructure.

Seed Distribution and Marketing: Develop efficient marketing channels and distribution networks to ensure timely and adequate availability of seeds for farmers.

Infrastructure Facilities: Invest in storage facilities, transportation infrastructure, and communication networks to support smooth seed delivery and market development.

Transgenic Plant Varieties: Establish regulatory frameworks for transgenic plant varieties while upholding biosafety concerns.

Import and Export of Seeds: Streamline seed import and export procedures while protecting plant health and preventing introduction of unwanted pests and diseases.

Promotion of Domestic Seed Industry: Provide incentives and support to boost the domestic seed industry and reduce dependence on imported seeds.

Strengthening of Monitoring System: Implement robust monitoring mechanisms to ensure effective implementation of the policy and address emerging challenges.

Impact

The policy has led to significant progress in the Indian seed sector, including:

Improved Access to High-Quality Seeds: There has been a substantial rise in the area dedicated to certified seed production, resulting in greater availability of quality seeds for farmers.

Greater Involvement of the Private Sector: The policy has encouraged private sector investment in research and development, accelerating the creation and adoption of improved crop varieties.

Improved Seed Quality: Strengthening of quality control mechanisms has improved the overall quality of seeds available to farmers.

Growth of Domestic Seed Industry: The policy has contributed to the growth of the domestic seed industry, reducing dependence on imported seeds.

Plant Quarantine Order (2003)

The Plant Quarantine Order (Regulation of Import into India), 2003, is a crucial regulation implemented by the Department of Agriculture & Cooperation in India. It aims to:

1. Prevent the introduction and spread of pests and diseases of plants: These include insects, pathogens, and weeds that could harm plant health and threaten biodiversity.
2. Regulate the import of plants and plant products: This guarantees that only disease-free and pest-free plant materials are allowed entry into the country.
3. Facilitate safe international trade in plants: By complying with international standards and agreements, the order promotes smooth trade while safeguarding plant health.

Key Features of the Order

Import Permit System: A permit is required for importing most plant materials into India. The permit application process involves inspections and pest risk assessments.

Banned and Regulated Items: Specific plants and plant-based products are entirely prohibited from being imported because they pose a significant threat of introducing harmful pests and diseases. Others are restricted and require specific quarantine treatments or certifications.

Post-Entry Quarantine: Imported plants may be subjected to further inspections and quarantine measures to ensure they are free of pests and diseases.

Inspection and Certification: Authorized phytosanitary inspectors at notified points of entry examine imported plants and issue phytosanitary certificates verifying their health status.

Penalties for Violation: Non-compliance with the order can result in penalties, including seizure and destruction of infected plant materials.

Benefits of the Order

Protects Plant Health and Biodiversity: The order safeguards domestic agriculture and natural ecosystems by preventing the introduction of harmful pests and diseases.

Promotes Sustainable Agriculture: Healthy plants contribute to higher yields, food security, and rural livelihoods.

Facilitates Trade: By complying with international standards, the order simplifies and fosters safe trade in plant materials.

Consumer Protection: Healthy imported plants and products ensure quality and safety for consumers.

Challenges and Future Perspectives

Capacity Building: Ensuring adequate infrastructure, trained personnel, and resources for effective enforcement is crucial.

Alignment with Global Norms: Regular revision and updates of the order are crucial to ensure it remains in line with changing international standards.

Stakeholder Education: It is vital to increase awareness among all relevant stakeholders—such as importers, traders, and farmers—regarding the provisions outlined in the order.

International Cooperation: Enhancing partnerships with other countries for conducting risk assessments, sharing information, and coordinating pest control strategies can significantly improve efforts to safeguard plant health.

Import Regulation of GM Products under the Foreign Trade Policy (2006): The regulation of genetically modified (GM) product imports in India, as outlined in the Foreign Trade Policy (2006), is intricate and constantly evolving due to the potential risks and advantages associated with GMOs. The process of bringing GM food, feed, or living modified organisms (LMOs) into India is strictly regulated and involves the following key components:

Licensing Protocols

- **Mandatory Prior Approval:** All imports of GM food items, feeds, organisms, or LMOs must first receive clearance from the Genetic Engineering Appraisal Committee (GEAC), which functions under the Ministry of Environment, Forest and Climate Change (MoEFCC).
- **Customs Release:** A No Objection Certificate (NOC) from the Food Safety and Standards Authority of India (FSSAI) is required before customs authorities can release the shipment into the country.

Additional Regulatory Frameworks

Environment Protection Act (1986): Ensures environmental safety considerations are addressed during import and use of GM products.

Food Safety and Standards Act (2006): Governs the import, production, storage, marketing, and distribution of food products, including those that are genetically modified (GM).

Rules for Import and Export of Genetically Modified Plant (2008): Provides specific procedures for import of GM plant material for research or experimental purposes.

Key Considerations

Risk Assessment: GEAC conducts a thorough risk assessment of each GM product, considering potential environmental and health impacts before granting approval.

Labeling Requirements: Imported GM food products must comply with FSSAI labeling regulations, clearly mentioning the presence of GM content.

Tracking and traceability mechanisms for imported genetically modified (GM) products across the entire supply chain are essential to ensure safety and adherence to regulatory requirements.

General Notes Regarding Import Policy

The import of any genetically modified (GM) food, feed, raw or processed material, or food product containing GM components must receive prior authorization from the Genetic Engineering Approval Committee (GEAC), which operates under the Ministry of Environment, Forest and Climate Change (MoEFCC).

The import consignment must be accompanied by a declaration stating whether it contains GM material. Failure to do so can lead to penalties under the Foreign Trade (Development and Regulation) Act, 1992.

Specific Regulatory Framework

The GEAC follows a risk assessment process to evaluate the safety of GM products before granting import approval. This includes considerations like potential environmental impact, human health risks, and socioeconomic implications.

The GEAC may impose specific conditions on import, such as labelling requirements, restricted use for specific purposes, or post-market monitoring.

Depending on the specific type of genetically modified (GM) product, other regulations may also be applicable—for instance, food safety regulations outlined under the Food Safety and Standards Act (FSSAI) of 2006.

National Environment Policy (2006)

The National Environment Policy (NEP) 2006 serves as a blueprint for sustainable development in India. It outlines the government's approach to environmental protection and conservation, aiming to balance environmental needs with economic and social development. Here's a closer look at its key objectives, principles, and impact:

Objectives

Conservation of Critical Environmental Resources: Protect and preserve vital ecosystems, natural resources, and biodiversity.

Prevention and Control of Environmental Pollution: Reduce and manage pollution across air, water, land, and noise to ensure public health and environmental quality.

Sustainable Development: Promote development activities that meet the needs of the present without compromising the ability of future generations to meet their own needs.

Environmental Governance: Strengthen institutional frameworks and public participation in environmental decision-making.

Equity and Justice: Ensure equitable access to environmental resources and benefits, addressing environmental concerns of vulnerable communities.

Key Principles

Precautionary Principle: Take precautionary measures to prevent environmental harm, even in the absence of conclusive scientific evidence.

Polluter Pays Principle: Hold polluters accountable for the environmental costs of their activities. Intergenerational Equity: Ensure that present actions do not compromise the ability of future generations to meet their environmental needs.

Public Participation: Encourage public participation in environmental decision-making and awareness-raising efforts.

Sustainable Use of Resources: Promote sustainable use of natural resources and efficient utilization of resources.

Impact

The NEP 2006 has had significant impacts on environmental policy and action in India, including:

Strengthened Environmental Legislation: The policy has spurred the development and strengthening of environmental laws and regulations across various sectors.

Greater Funding for Environmental Safeguards: There has been a rise in financial support for environmental projects, resulting in better pollution control systems and stronger conservation measures.

Improved Community Engagement: Awareness of environmental concerns has grown, leading to more active involvement of the public in policy-making and environmental decision processes.

Shift towards Sustainable Development: The policy has fostered a shift towards sustainable development principles in various sectors, such as energy, agriculture, and urban planning.

3

Technology Commercialisation

Technology Assessment and Refinement

What is Technology Assessment?

Technology assessment serves as a method or tool for forecasting future technological needs. It helps in envisioning or predicting potential technological developments required by users and in identifying areas where existing technologies can be improved. In agriculture, evaluating a technology means understanding its worth or effectiveness. This evaluation can be done by researchers, farmers, or other relevant stakeholders. The knowledge obtained from such evaluations is essential for creating new technologies, improving existing ones, and ensuring their successful adoption by end users.

The approach to technology assessment varies depending on who is conducting the evaluation, the purpose behind it, and the specific characteristics of the technology being assessed. It is commonly defined as the organized and intentional use of knowledge, skills, and experience to carry out a task or deliver a service that contributes to the well-being of society.

Technology Assessment: Definition and Objectives

According to Strasser (1972), technology assessment is a structured approach to planning and forecasting that identifies various options and associated costs. It considers not only economic factors but also environmental and social impacts, both internal and external to the technology being evaluated. The assessment places emphasis on analyzing both the positive and negative consequences of technologies. As such, institutions involved in the development, refinement, and dissemination of agricultural technologies have a significant stake in conducting technology assessments.

Objectives of Technology Assessment

1. To enhance the utilization of existing and upcoming technologies.
2. To address and adapt to new and emerging challenges.

3. To inform and support policy decisions related to technology, including those linked to international agreements such as GATT and WTO.
4. To establish frameworks and policies that promote technologies with greater benefits and fewer negative impacts.
5. To innovate, improve, and share technologies that effectively address the societal and practical issues faced by farmers.

Technology assessment involves a comprehensive and impartial evaluation of all key impacts direct, indirect, immediate, or long-term of a technological innovation on society, the environment, or the economy. This process aims to align science and technology with human needs by selecting the most appropriate technologies to improve quality of life from good to excellent. The growing significance of technology assessment lies in the fact that not all technological changes lead to actual progress or better living standards.

How is the Technology Evaluated?

The concept of technology assessment originated with the U.S. Navy, where it was used to evaluate arms and equipment based on their usefulness to sailors in combat. From this, technology forecasting emerged as a strategic management tool to anticipate future needs and guide the development of relevant technologies (Cetron, 1971). Technology is evaluated using a set of criteria, indicators, or parameters, depending on the context.

For example, when evaluating onions for export, aspects such as size, taste, yield, and color are considered. In the case of flowers, important traits include spike length, color, and scent. For processed products, the focus is primarily on value addition and consumer preferences. Therefore, technology assessment is highly context-dependent and differs based on the type of technology, the intended users, and the specific environment in which it is applied.

Each criterion used for assessment is assigned a level of relative importance. Technologies are then scored on a 10-point scale based on their positive or negative impact on each criterion. The final score for each indicator is derived from multiplying the weight of the criterion with the consequence score, the total of all the indicators provides the overall assessment of the technology's value for the targeted user.

Why is Technology Assessment Attaining Importance?

Technology assessment has become increasingly vital, especially in agriculture. The technological base expanded significantly during the Green Revolution in the mid-1960s, which saw a major boost in productivity and national food production. While numerous technologies released from research institutions showed high potential in controlled conditions, their benefits often reached only a limited segment of farmers, leaving many others behind. At present, close to one billion individuals employed in agriculture worldwide operate on land holdings of less than one hectare. The liberalized trade environment post-GATT has made global agriculture more specialized and competitive. For countries like India, where land is scarce and rural labor is abundant, diversification towards high-value crops is essential. This requires intensive, precision-based farming systems that can only be developed after careful technology assessment.

With the privatization of agricultural research and extension, public funding is being reduced. This shift tends to favor wealthier, medium- and large-scale commercial farmers who have the technical knowledge, financial capacity, and resources to adopt advanced technologies. As a result, smallholder and marginal farmers risk being excluded. Therefore, it's essential to evaluate technologies for specific micro-environments and offer a range of suitable alternatives to end users.

In the 1960s, India's national agricultural extension strategies assumed that Western technologies could be directly applied without adaptation. These were not evaluated or adapted to suit local conditions. Although rice and wheat varieties introduced from IRRI and CIMMYT showed potential, their adoption was limited because of the high costs associated with seed replacement (Dalrymple and Srivastava, 1994).

Subsequently, on-farm adaptive research was implemented to provide recommendations specific to India's 126 agro-climatic zones. However, these evaluations were generally carried out by scientists rather than involving farmers directly. Today, the primary challenge for public research is to create efficient and sustainable technologies that cater to nearly one billion resource-poor farmers working in complex, diverse, and risk-laden agricultural settings commonly known as CDR agriculture.

Thus, precision technologies tailored for these farmers—who constitute about 66% of India's farming population—are urgently needed. Singh (1996) emphasized that a technology's appropriateness is crucial for its adoption. A highly effective method for evaluating this is by conducting a cost-benefit analysis that takes into account inputs such as physical and biological resources, environmental effects, potential risks, and socio-economic factors.

The Institution–Village Linkage Programme (IVLP), launched under the National Agricultural Technology Project (NATP), aimed to ensure stronger collaboration between scientists and farmers using a bottom-up approach. Implemented in 295 villages across 70 centers, the program involved over 37,000 farmers and led to 1,671 technology interventions across various sectors—crops (923), livestock (247), horticulture (306), forestry (12), fisheries (37), gender issues (53), and others (93).

In 2010, Krishi Vigyan Kendras (KVKs) evaluated 1,913 technologies at 6,574 locations through 18,425 field trials on 465 crops. These assessments focused on areas like varietal performance, nutrient and pest management, cropping systems, resource conservation, value addition, and entrepreneurship. For animal sciences, 189 technologies were assessed across 877 locations with 2,692 trials, covering animal health, breeds, nutrition, and production practices. Furthermore, 235 technologies were refined at 803 sites through 4,911 field trials. KVKs also evaluated 99 technologies in 280 locations involving 1,025 rural women, focusing on themes like drudgery reduction, health, nutrition, and entrepreneurship development.

A technology refers to any product, process, or method used to achieve a specific desired outcome. A critical technology specifically addresses and helps solve pressing issues faced by users within their specific context. Technology assessment involves evaluating the value of a technology based on its potential, available resources, opportunities, and the social and economic factors surrounding it—both internal and external. This evaluation emphasizes the technology's benefits and drawbacks in relation to the user's situation (Cetron, 1971). The indicators used in such assessments differ depending on the region, the target groups, available resources, and the specific context. According to the Brundtland Commission, India's target groups can be categorized into three: farmers practicing industrial agriculture, those engaged in green revolution agriculture (whether fully or partially irrigated), and those involved in resource-poor farming systems (Misra, 1993).

Can the Farmer Assess Technology?

Yes, farmers can play an active role in technology assessment using participatory approaches such as Participatory Rural Appraisal (PRA). Studies have found that informal methods and farmer-led technology evaluations are not only feasible but also effective. In programs like the Institute Village Linkage Programme (IVLP), farmers have worked alongside multidisciplinary teams to assess and evaluate technologies collaboratively.

These assessments were guided by Rural People's Knowledge (RPKs), allowing for categorization of ecosystems into favourable and unfavourable zones. Based on these assessments, technology recommendations are developed specifically for particular regions and user groups.

Impacts of Technology Assessment on the Process of Technology Transfer

India's current self-sufficiency and market surplus in agriculture are largely due to the Green Revolution, which focused on area-based transfer of technology packages, emphasizing production potential. While initiatives like IADP (Intensive Agricultural District Programme), IAAP, and HYVP achieved production goals, they often excluded small and marginal farmers. To bridge this gap, initiatives like SFDA, MFAL, and IRDP were launched to foster development that upholds social justice and encourages inclusive participation.

To further refine agricultural technologies, the Indian Council of Agricultural Research (ICAR) institutionalized the process of technology assessment and refinement in 1995 through the IVLP. Launched during the rabi season of 1995–96, the program initially operated in 42 centers nationwide, reaching 42,000 farm families. As part of the National Agricultural Technology Project (NATP), it aimed to scale up its reach to 200,000 farm families, supported by a financial allocation of ₹ 700 million.

One of the key objectives of IVLP, particularly objective number 7, is to identify extrapolation domains geographical areas where new technologies or technology modules can be extended based on environmental characterization at both meso and macro levels. These extrapolation domains are determined through technology assessments that focus on selected indicators, identifying gaps in current technologies and modifying or developing new ones tailored to specific user needs. This participatory approach empowers farmers to play a more active role in their development.

For example, in Mali, farmers assessed different maize varieties by either accepting, rejecting, or modifying the technologies. The factors influencing their decisions were determined through informal surveys and analysis of existing data sources.

It is essential that technology assessments concentrate on the real-world and technical conditions in which the innovations will be applied, instead of depending solely on biological performance data from research stations. The selection of assessment methods, indicators, and variables should be guided by the specific local context and the resources at hand.

A study conducted in the Philippines on technology assessment in mango cultivation applied several criteria, including technical feasibility, economic viability, social and cultural acceptability, environmental safety, sustainability, ease of application, clarity of understanding, and efficient use of resources. One of the key benefits observed from using chemical sprays to induce off-season flowering was the ability to disrupt the typical seasonal supply pattern and address the issue of alternate bearing in mangoes. However, a major drawback was that prolonged use of these flowering inducers eventually led to a reduction in mango yields over time (Granola, Jr. et al., 1988).

When technology assessments incorporate both qualitative and quantitative data along with multiple indicators, tools such as scoring models, checklists, and logical frameworks have been effectively utilized (McClean, 1988; Nestel, 1989).

Issues for Technology Assessment

Before carrying out a technology assessment, it is important to consider several essential research questions:

1. What particular indicators do farmers rely on to assess the usefulness and suitability of different technologies, and how do these indicators affect their decisions to adopt, reject, or modify them?
2. Do these evaluation indicators differ between large-scale, small-scale, and marginal farmers?
3. If such variations are present, what approaches can be implemented to harmonize technology assessment criteria among these different farming groups?

Technology assessment plays a crucial role in identifying research gaps. It examines whether appropriate recommendations exist for a district's specific issues under each Agro-Ecological Situation (AES). Similarly, extension gaps are identified by analyzing the discrepancy between recommended practices and their actual adoption. This is done by understanding the challenges farmers face, using participatory methods.

SWOT analysis plays a key role in the Strategic Research and Extension Plan (SREP) by assessing the strengths and weaknesses of existing farming systems. It also helps in identifying opportunities for improved resource utilization, access to new markets and technologies, and recognizing potential threats to natural resources and market stability. Furthermore, recording the achievements of progressive and innovative farmers serves as a source of motivation for others, promoting quicker adoption of technologies and enhancing problem-solving within the farming community.

Steps for Conducting Technology Assessment

As outlined in the Institute Village Linkage Programme (IVLP) by ICAR (1995), the following steps are followed for technology assessment:

i) Identifying constraints within the farming system using methods like Rapid Rural Appraisal (RRA) or Participatory Rural Appraisal (PRA);

ii) Selecting innovations or interventions that specifically address the identified constraint—these are referred to as critical technologies or key inputs;

iii) Determining the set of criteria or indicators that define the farming system and are relevant for evaluating the effectiveness of the technologies in addressing the targeted issue;

iv) Comparing all identified technologies based on these selected criteria; and

v) Validating the assessment process by sharing it with subject matter experts and gathering their feedback.

Sampling Farmers for Technology Assessment

To carry out a technology assessment, the process can begin by selecting a group of farmers through Participatory Rural Appraisal (PRA). This approach helps in identifying key informants who are well-positioned to propose meaningful and context-specific indicators for evaluating the technology.

Creating an Indicator Pool for Technology Assessment

A preliminary list of approximately 150 potential indicators can be compiled with input from experts, field observations, and consultations with farmers. These items should relate specifically to the area of technology application—such as crop production. Farmers can then rate each item on a 10-point scale, ranging from "least important" to "most important."

Calculating Scale Values

The relative significance (weights) of each indicator is determined using the normalized ranking method. Here's the procedure:

1. Rank the indicators based on their total scores.
2. Convert these ranks to percentile values using the formula:

   $P=(Ri-0.5)n\times 100$

   Where:

   - Ri = Rank of the *i*th indicator
   - n = Total number of indicators
3. Convert percentiles to z-scores (normalized values).
4. Use a linear transformation to convert z-scores into scale values (C) with a mean of 5.00 and standard deviation of 2.00:

   $Z=C-5.00/2.00$

Constructing the Technology Assessment Index (TAI)

Two approaches can be used to build the index:

- Unweighted composite: Treats all indicators equally.
- Weighted composite: Assigns different weights to each indicator based on their scale values.

The composite score is computed as: $Xc=a1I1+a2I2+\ldots+anIn$

Where

- Xc = Composite index score
- a1…an = Weights (from scale values or factor loading)
- I1…In = Individual indicator scores

Indicators can be positively or negatively worded. The raw scores range from +1 to +5 (positive impact) and –1 to –5 (negative impact), based on the respondent's rating.

Steps to Compute Technology Assessment Index Score (TAI)

1. Multiply the raw score of each respondent on every indicator by its respective scale value.
2. Convert the product into unit scores using:

$U_{ij} = y_{ij} - \text{Min } y_j / \text{Max } y_j - \text{Min } y_j$

Where:

- y_{ij} = Raw score of the *i*th respondent on the *j*th indicator
- Min y_j, Max y_j= Minimum and maximum possible scores on the *j*th indicator
- U_{ij}= Unit score ranging between 0 and 1

3. Sum the weights of the scale values used.
4. Calculate the final Technology Assessment Index (TAI) score using:

   $TAI = \sum U_{ij} / \sum a1 \ldots an \times 100$

The resulting TAI score ranges between 0 to 100.

Score Interpretation

Because raw scores include both positive and negative values (after being multiplied by scale values and standardized), the final TAI is expressed as a percentage.

- TAI < 50 indicates a net negative impact.
- TAI ≥ 50 indicates a net positive impact.

To avoid clustering most scores in the "medium" category due to variations in mean and standard deviation across different technologies and farmer groups, a fixed interpretation scheme is used:

This classification is tailored to each client (or group of similar clients), and the average TAI for a category of farmers can be used to infer the overall effectiveness of a technology for that group. It also aids in customizing technology recommendations for specific farming contexts.

Category or Index score	**Technology effect**
>75	Highly positive
>50 up to 75	Positive
<50 up to 25	Negative
<25	Highly negative

Participatory Technology Assessment

Participatory technology assessment (PTA) has been developed as a way to "engage a representative group of lay people in the processes of science and technology decision- making".

Participatory technology assessment (PTA) states that a class of methods to integrate new voices into science policy discussions which encompasses various forms of public deliberation including citizens' assemblies, citizens' juries, and consensus conferences.

- **Aim**

The purpose of 'participatory technology assessment' is to offer guidance to policymakers and to foster broader public discussion on social and technological advancements.

- **Methods and Approaches**

The Participatory Technology Assessment method includes three phases of participatory activity:

a) **Problem Framing**: This first stage consists of open-framing focus groups with 15-20 selected citizens and a stakeholder design workshop. The focus groups elicit citizen perspectives on an issue with minimal or no background material to limit expert issue framing.

b) **ECAST Deliberation**: Based on the World-Wide Views deliberation method developed by the Danish Board of Technology, the Expert and Citizen Assessment of Science and Technology (ECAST) involves full-day, informed discussions with 80 to 100 participants from the general public.

c) **Results Integration**: This last stage is the analysis of both qualitative and quantitative data collected during the event. After a preliminary analysis of the data, a second stakeholder workshop is convened to present preliminary deliberation results and solicit feedback on directions for deeper analysis. Since we collect more data than we can analyze in their entirety, we host a preliminary results workshop to solicit expert and stakeholder input on areas for deeper analysis. This process helps to ensure the outputs of our deliberations are useful to policy and decision makers.

Technology Valuation Returns to Investment

Understanding Assets

An asset refers to any resource owned or controlled by an individual or business due to past actions (such as purchase or internal development) that is expected to provide economic benefits in the future—either by generating cash inflows or reducing costs.

A business's total wealth can generally be divided into three types of assets:

Wealth=Working Capital + Fixed Assets + Intangible Assets

Types of Assets

1. Working Capital

- It is the surplus of a company's current assets (like cash, short-term investments, inventory, and receivables) over its current liabilities (such as payables, short-term debts, and taxes).
- Also called net current assets, it represents the day-to-day operational liquidity of a business.

2. Fixed Assets

- These are long-term tangible resources used in business operations, such as land, buildings, machinery, vehicles, office equipment, and computers.
- Fixed assets are not typically sold or converted into cash during normal operations and have traditionally been seen as the backbone of a company's value.

3. Intangible Assets

- These include non-physical properties of a business.
- Historically grouped under "Goodwill", these may include customer loyalty, brand reputation, employee morale, and intellectual property.
- Goodwill refers to the amount paid for a business that exceeds the fair market value of its identifiable net assets.

Intellectual Property (IP) Assets

IP assets are a distinct type of intangible assets that receive legal recognition and protection. Their uniqueness lies in their creation through law, which makes them enforceable, transferable, and identifiable. While their legal life may be long, their economic usefulness (economic life) is usually shorter.

Examples of IP assets

- Patents
- Trademarks
- Copyrights

- Industrial Designs
- Trade Secrets

Perspectives on IP Assets

- Legal View: Defined based on legal criteria like novelty or originality.
- Economic View: Defined by their ability to create economic value.

Example: A legally valid patent that doesn't contribute to income has no economic value despite its legal status.

Price and Value

- Price is the amount for which an asset is bought or sold in the market.
- Value refers to the expected economic gains the owner or user will receive from the asset in the future.

Valuation involves estimating the financial worth of an asset. While sometimes the market price may reflect an asset's value, this is not always the case.

Valuing Assets

The worth of any asset—whether it's a physical item like machinery or an intangible one like intellectual property—depends on the economic benefits it is expected to generate in the future. While some assets are simpler to assess, the accuracy of valuations can vary significantly.

Valuing IP Assets

The worth of an IP asset is based on its exclusive use, granted by law. This exclusivity allows businesses to prevent others from using or copying it—giving them a competitive edge.

To have measurable value, an IP asset must

- Generate direct and quantifiable economic returns.
- Improve the value of other associated business assets. Ways to Extract Value from IP

1. Direct Use
   - The company itself uses the IP to create and sell products or services.
2. Licensing or Selling
   - The IP can be licensed out or sold to generate revenue.

3. Strategic Ownership Without Use

Even unused IP can add value by:

- Limiting customer negotiation power
- Diminishing the power of suppliers
- Lessening the intensity of competition
- Increasing obstacles for potential new entrants
- Reducing the risk posed by substitute products

IP Valuation Triggers

There are various specific reasons or motivations for carrying out an intellectual property (IP) valuation.

These are known as valuation triggers, which indicate the underlying purpose or need for conducting the valuation. Some of the common triggers include:

Category	**Situation triggering Valuation**
Transaction	Licensing or franchising of intellectual property (IP); buying or selling IP assets; mergers and acquisitions; forming joint ventures or strategic partnerships; donating IP assets.
IP rights Enforcement	Estimating compensation or damages in cases of IP infringement.
Internal Application	Making decisions on R&D investments; managing IP assets internally; strategic funding or capital raising; maintaining investor relations.
Other uses	Financial disclosures and reporting; bankruptcy or asset liquidation; tax planning and optimization; insuring IP assets.

When IP Valuation is Done?

Valuation of intellectual property (IP) plays a critical role in various business and legal scenarios, such as:

- **Litigation** – used in legal disputes over IP rights.
- **IP Audits** – a strategic tool to evaluate and manage IP assets.
- **Licensing** – helps determine royalty rates and contract terms.
- **Joint Ventures & Collaborations** – useful for fair valuation in partnerships.
- **Mergers and Acquisitions** – supports due diligence and pricing of IP portfolios.

- **Financial Reporting** – aligns with accounting and disclosure practices.
- **Securing Financing** – assists in using IP as collateral or a valuation benchmark.
- **Investment Transactions** – aids in negotiating deals or attracting investors.

IP Valuation in Research & Development (R&D) Settings

In R&D-focused environments, IP valuation serves as a valuable strategic and operational tool. It helps to:

- **Support decision-making** during both pre-commercial and commercialization phases.
- **Facilitate technology transfer** through licensing, partnerships, or spin-offs.
- **Enable fundraising** by showcasing the value of developed technologies.
- **Effectively convey** the importance of innovations to all relevant stakeholders.
- **Enhance learning** by revealing ways to strengthen processes, research output, and human capital.
- **Track returns on investments (ROI)** in innovation.
- **Assist in legal matters**, although litigation is rare for publicly funded institutions.

IP Valuation Approaches

IP valuation typically involves **both qualitative and quantitative** components:

1. Qualitative Valuation

Qualitative assessment helps set the foundation for financial models by analyzing various factors that influence IP value, such as:

- The **intrinsic quality** of the intangible asset.
- Its **strategic role** in the business relative to other value drivers.
- The **industry context** and **competitive landscape**.
- The **economic environment** over the asset's useful life.

Key Components

- **Scoring criteria**: What aspects are measured.
- **Scoring system**: How those aspects are rated.
- **Scoring scale**: The range of scores used.
- **Valuation factors**: Influential characteristics or conditions.
- **Decision rules**: How scores lead to value judgments.

2. Quantitative Valuation

Quantitative methods estimate the **monetary worth** of an IP asset and include:

- **Cost Approach**
- **Market Approach**
- **Income Approach**

Cost-Based Valuation

This approach estimates the value of intellectual property by considering the costs involved in its creation or replacement, taking into account both direct expenditures and opportunity costs. It is typically used when:

- The IP is **not currently generating revenue**.
- The valuation is required for **internal purposes**, such as budgeting or project planning.

Steps in Cost Method

- Assess historical development costs.
- Adjust for **inflation** to find current cost equivalents.
- Factor in **obsolescence** to reflect depreciation in value.

Types of Obsolescence Affecting IP

Although IP does not physically deteriorate, its value may reduce due to:

- **Functional Obsolescence**: When using the IP incurs higher operational costs compared to newer alternatives.
- **Technological Obsolescence**: When newer technologies make the existing IP outdated or irrelevant (e.g., outdated patent on floppy disks).

- **Economic Obsolescence**: When the IP cannot generate sufficient returns in its best possible application.

Note: Cost-based valuation is **least preferred** among the methods for IP appraisal, as it often does not imitate the true market value. It is mainly used as a supplementary method, particularly when the IP is not actively producing income.

The cost method has two main variations: the reproduction cost method and the replacement cost method.

Reproduction cost method	Replacement cost method
- This method estimates the cost of creating an identical copy of the existing intellectual property (IP). - It calculates the total expense, based on current prices, to produce an exact replica of the IP asset. - The duplicate would be made using materials, standards, designs, layouts, and quality similar or identical to those used in the original asset. - However, this method does not consider technological advancements, improved materials, or enhanced utility that might now be available.	- This approach estimates the cost of developing a new asset that performs the same function or offers the same utility as the original IP, though it may look or function differently. - It determines the total cost, at current prices, to create an asset that delivers equivalent usefulness or performance. - The replacement asset may offer improved functionality or greater utility compared to the original. - It is typically developed using modern techniques, up-to-date standards, advanced design, and cutting-edge technology. - If the replacement asset is superior in any way, the added value or satisfaction it provides must be factored into the calculation, especially when considering obsolescence.

A key requirement when using cost-based valuation methods is that **all costs must be assessed as of the valuation date**—not based on past expenditures. This is because:

- Some resources used in developing the asset may now be publicly available, reducing current costs.
- Advances in technology or methods may allow similar results with less time or effort, lowering development costs today.

Reproduction Cost vs. Replacement Cost Methods

Overview of the Cost Method:

Step 1: Calculate Replacement Cost

Reproduction Cost - Curable Functional and Technological Obsolescence =

Replacement Cost

Curable obsolescence refers to deficiencies that can be corrected economically—i.e., the benefit of fixing them is more than the cost to do so.

Step 2: Estimate IP Value

Replacement Cost - Economic obsolescence - Incurable functional and technological obsolescence = Value

- Incurable obsolescence includes flaws that are too expensive to correct, with benefits not justifying the cost.

When to Use Each Method: Reproduction Cost Method

Used when the goal is to determine the cost of creating an exact duplicate of the asset. Suitable for:

- Legal proceedings (e.g., infringement cases)
- Calculating Return on Investment (ROI)
- Tax assessments (especially for embedded software)

Replacement Cost Method

Used to estimate the cost of building a functionally equivalent asset using current technologies and practices. Appropriate for:

- Setting a price for buying or selling an IP asset
- Determining royalty bases
- Transfer pricing decisions
- Valuing outdated brands in today's market (e.g., adapting an old brand for modern platforms like online or podcasts)

Choice of method depends on:

1. The nature of the IP asset,
2. The valuation date,
3. The context or purpose of valuation.

Market-Based Valuation Method

This method compares the target IP asset with **similar IP assets that were recently bought or licensed** in independent, arm's-length transactions.

Key Points

- Relies on **real transaction data** such as purchase price or royalty rates.
- Works best when there is **comprehensive data** about the IP and transaction terms (like whether it involved litigation settlement or cross-licensing).
- Reflects current **market trends and sentiments** more accurately than income-based models.

Income-Based Valuation Method

Also known as the economic benefit approach, this method estimates the value of an intellectual property (IP) asset by calculating the future income it is expected to generate, and then discounting that income to determine its present value.

Steps in this method

1. **Forecast future income** (or cost savings) expected from the IP over its remaining useful life.
2. **Deduct related expenses**, such as labor, materials, capital costs, and any economic charges.
3. **Apply a discount rate** to convert the projected income stream to present-day value, accounting for risk and time.

This is the most broadly used process for IP valuation.

Types of Income-Based Approaches

1. Direct Capitalization

- Estimate the income expected in a single future period and divide it by a chosen **capitalization rate** (i.e., expected rate of return).
- Useful for stable, predictable income streams.

2. Discounted Cash Flow (DCF)

- Projects detailed income over multiple periods and discounts each to present value.

3. Monte Carlo Simulation

- A probabilistic model that assesses value under different scenarios and uncertainties.

4. Real Options Method

- Values IP considering flexible decision-making under uncertainty, much like financial options.

5. Rule of Thumb

- Quick estimation using industry averages or benchmarks (e.g., % of sales).

6. Relief from Royalty Method

- Calculates value based on hypothetical royalties the business would save by owning the IP instead of licensing it.

Discounted Cash Flow Approach (DCF)

- (Discounted future economic benefits)
- The valuer projects the appropriate measure of economic income (cash flows) for
- several discrete time periods into the future.
- This projection of prospective economic income (cash flows) is converted into a
- present value by the use of a present value discount rate.
- DCF is the most frequently used approach of the Income Method.
- It involves estimating the future net cash flows anticipated from the commercial exploitation of the intangible asset being evaluated.
- This projection spans the entire economic lifespan of the intellectual property (IP).
- The expected cash flows are adjusted using a discount rate to account for the time value of money and associated risks.
- The main goal is to calculate the Net Present Value (NPV) of the IP asset.

$$DCF = \frac{CF_1}{(1+r)^1} + \frac{CF_2}{(1+r)^2} + \ldots + \frac{CF_n}{(1+r)^n}$$

DCF = Discounted Cash Flow

CF = Cash Flow period i r = Interest rate

n = Time in years before the future cash flow occurs

Monte Carlo Simulation

- The Monte Carlo Simulation is an advanced, computer-assisted enhancement of the Discounted Cash Flow (DCF) method that incorporates multiple scenarios.
- Rather than relying on fixed input values, this method assigns a range of possible values to each component in the DCF model, using different statistical distributions such as normal or triangular.
- It generates thousands of simulations to project possible outcomes for net present value (NPV), displaying the results in a frequency histogram that visually shows the likelihood of different valuation results.
- This approach helps assess risk and uncertainty in a more detailed and probabilistic way.

Real Option Method

- The Real Option valuation technique is inspired by the Black-Scholes model (1972), which was originally developed to value financial options—the right to buy or sell an asset at a fixed price before a set date.
- This method draws a parallel between financial options and intellectual property, as both involve uncertain future benefits and decisions under uncertainty.
- It follows income-based valuation principles, especially the Discounted Cash Flow (DCF) method, but also includes the idea of strategic flexibility—such as the ability to delay, expand, or abandon a project.
- It is mathematically complex and relies on difficult-to-estimate inputs such as:
  - The current value of the asset (used as NPV),
  - The volatility or variability of the asset's future value,
  - The expected cash flows from the asset (comparable to dividends),
  - The strike price (i.e., the investment needed to commercialize the technology),

  - The economic life (analogous to the option's expiration date),
  - The risk-free interest rate (typically a long-term government bond rate).

- Real option valuation is especially valuable for assessing early-stage technologies, where future developments are highly unpredictable and decision-making can adapt as new information becomes available.

"Rule of Thumb" Method

- Under the Rule of Thumb approach, a common licensing practice assumes that the licensor (IP creator) receives 25% to 33% of the licensee's profits, not revenues.
- While simple and widely referenced, this method is controversial due to its lack of precision and contextual factors.
- It is often used as a rough benchmark or initial indicator, especially when detailed data or methods are unavailable.
- However, U.S. courts have formally prohibited its use in litigation due to its non- rigorous, formulaic nature.

Relief from Royalty Method

- The Relief from Royalty Method is also referred to as the Royalty Avoidance Method.
- It is one of the business valuation methods under the income approach which is used for the valuation of intangible assets such as brand names, franchises, and trademarks.
- This method operates on the assumption that the asset's value to the buyer lies solely in avoiding the need to pay royalties for its use.
- There are databases in Organisation of Economic Co-operation and Development (OECD), Canada, USA and other jurisdictions that regularly publish prevailing royalties across industries and business sectors for brands, trademarks, franchises, etc which offer comparable for the intangible asset being valued.
- The relief from royalty method determines the fair market value of an intangible asset by calculating the present value of the royalty payments that the asset's owner avoids or is "relieved" from paying, as they hold the rights to use the asset themselves.

- First, the analysts identify the royalty percentage which is applied to the forecast revenue.
- Second, the analysts compute the discount rate.
- Third, the analyst discounts the royalties by the discount rate to arrive at the fair market value of the intangible asset.

Industrial Standard

Standard Industrial Royalties

- Certain industries have established typical royalty rates over time, often based on commonly accepted "rule of thumb" practices.
- Inconvenient for IP – patents and other IP aren't commodities and thus cannot be accurately valued at a set rate.
- However, if a patent is being valued for an external transaction within an industry that traditionally applies standard royalty rates, then the use of this standard rate in the valuation cannot be totally dismissed.
- For an internal valuation, the use of standard royalty rates is not recommended.

Technology Commercialisation Strategies

What is Technology Commercialization?

Technology commercialization is the process of taking an idea to market and creating financial value with it – typically through licensing an invention, developing a new product or service, or creating a new business. Products or services created through commercialization may be "new to the world" or just new to a region or country.

Technology commercialization does not necessarily refer to moving a specific "finished technology" to market, but may involve commercializing an earlier stage development. Sometimes technology commercialization involves marketing the know-how or intellectual property associated with a technology to companies that need what's "inside the black box", but do not need the box itself.

There are four overall stages in the technology transfer/ commercialization process:

a) **Invention Disclosure**: The first step in the invention review process is the formal disclosure of your invention. An Employee Invention Report describes the technology and is used to evaluate the commercial potential of the technology.

b) **Protection**: After determining whether your invention will make a viable product that will have a commercial partner, the next step in the process is protection of your invention by copyright, trademark, GI etc,

c) **Marketing**: It will help to identify potential commercial partners that have the resources and capacity to bring your technology to market.

d) **Licensing**: A license is an official permission or permit to do, use, or own something. A license is an agreement where owner grants another party permission to use the invention. After licensing, additional development for that technology to reach the market will be needed. Once a technology reaches the market, inventors are awarded a portion of royalties as set out in the licensing agreement.

Why is Commercialization Important?

Commercialization plays a huge part in deciding the role of an idea, product, or service in society. It generates job opportunities, creates choices for consumers, and paves the way for more market needs. It affects a nation's economy and sustains the quality of life standards. It also demands the learning and propagation of new skill sets, knowledge, and innovations.

Commercialization Strategy

It is the planned approach a company uses to secure funding and support to take its technology or product from the idea stage to the market. It involves selecting from various financial and strategic pathways to enable successful product launch and growth.

Some common and effective commercialization strategies used by advanced technology companies include:

- Licensing agreements that include development funding
- Forming strategic partnerships or alliances
- Securing equity investments in the parent organization
- Attracting equity investment in a newly formed spin-off company
- Launching an Initial Public Offering (IPO) to raise capital through public markets

Strategy 1: Licensing Accompanied by Development Funding

Many lifestyle firms favor licensing as their primary commercialization strategy, allowing them to focus on technology development while licensees manage marketing, sales, and manufacturing. As R&D budgets shrink in larger companies, licensing offers an appealing way for these firms to access new markets and maintain competitiveness without the high costs of original research and development.

Strategy 2: Strategic Alliances

These partnerships can include collaborations for technology testing, licensing, investment opportunities, and more. Common forms of these alliances are marketing collaborations, R&D efforts, manufacturing agreements, equity stakes, joint ventures, and licensing arrangements.

Strategy 3: Equity Investment in the Parent Company

In many cases, founders have a clear vision from the beginning about the kind of business they want to build. If the goal is to develop a high-growth, high-potential venture, the founder will make strategic decisions early on to lay the foundation for rapid expansion.

Strategy 4: Equity Investment in a Spin-Off

This is a more sophisticated approach that combines elements of previous strategies. Here, the founder chooses to maintain the core company as a small, privately owned entity focused on research and development. At the same time, they consider creating a separate spin-off company designed to attract equity investment and grow as a high-potential venture. For this model to work effectively, careful planning is needed to handle matters such as staffing, intellectual property rights, and non-compete agreements.

Strategy 5: The Initial Public Offering (IPO): An IPO is a prominent and often symbolic strategy for commercialization that highlights a company's success and provides opportunities for raising capital and allowing early investors to exit. While going public can boost visibility and open doors for mergers or acquisitions, it is also the most expensive route due to high costs associated with underwriting, legal services, and regulatory requirements. Public companies are also subject to increased scrutiny, with ongoing obligations to disclose information to a broad base of shareholders.

Approaches of Technology Commercialization

Scale-up an innovation

Here are three ways to effectively scale your organization's innovation strategy:

1. Build an expansive innovation network. Successful innovation strategy comes from building an empowered innovation network, both internally and externally
2. Develop strong processes and systems
3. Constantly evaluate success (or failure)

Licensing

Commercialization process ends with transferring intellectual property rights to a company prepared to bring technology to public use via a license. This license can be awarded to an existing company or a new startup formed by the original research team or others. Each project is unique, and licensing strategies are tailored to maximize benefits for both the university and the partner.

Option-to-license

Most often used to secure rights to a technology for a limited time frame. Options typically grant exclusive rights for a limited time frame to allow the partners a chance to evaluate the available opportunity and move forward with preliminary commercialization steps.

Exclusive License

Allows access to technology by a sole licensee and can be provided to either an established company or a startup company. This agreement provides exclusive rights, most often to a patented technology, in exchange for some financial consideration. Terms often include a royalty based on sales with some diligence provisions in future years such as milestone events and payments or annual minimum royalties.

Non-Exclusive License

Allows non-exclusive access to the technology by numerous industry partners. Terms typically include a royalty based on sales which are lower relative to exclusive license terms. Non-exclusive licenses do not often include diligence provisions but may include an agreement maintenance fee structure or other means of ensuring that the licensee actively pursues commercial implementation of the technology.

Technology Business incubator (TBI)

A Technology Business Incubator (TBI) aids tech startups by offering resources such as infrastructure, research assistance, funding, consulting, and marketing support. Their goal is to enhance economic development by improving the survival and growth rates of new ventures and transforming ideas into self-sustaining businesses. Over the past two decades, TBIs have expanded globally, helping countries leverage technology for economic growth.

Handholding

Hand holding" in technology commercialization refers to the support and guidance provided to innovators and entrepreneurs as they navigate the complex process of bringing new technologies to market.

Agri-preneur Development

- Agripreneurs are entrepreneurs focused on the agricultural sector, often using innovative approaches to improve farming practices, enhance productivity, and create sustainable businesses.
- This can include adopting new technologies, developing value-added products, or creating services that improve efficiency in the agricultural supply chain.
- The purposes of agricultural entrepreneurship are: Stabilizing market prices of agricultural commodities.
- Generating assured income from farm produce creating opportunities to get additional income by utilizing farm produce.
- Utilizing the additional revenue or surplus money to develop a viable business Generating adequate income to sustain farmers' livelihood.

Agri-Business Incubators (ABI)

Open new entry points in the agricultural value chains, which in turn keep in accessing new market. Incubators help by strengthening and facilitating linkages between enterprises and new commercial opportunities. Agri-Business incubation is defined as a process which focuses on nurturing innovative early-stage enterprises. These enterprises have high growth potential to become competitive agribusinesses by serving, adding value or linking to farm producers

Example: ABI-ICRISAT was established in 2003 by ICRISAT with support of the Department of Science and Technology (DST), Government of India, under its National Science & Technology Entrepreneurship Development Board (NSTEDB).

As the first agricultural sector focused incubator, ABI-ICRISAT is recognized as a Technology Business Incubator (TBI) that aims to promote agri-based entrepreneurship in the drylands, and facilitate public-private partnerships for growth. Nurturing ag/food tech start-ups towards developing technology-led innovative solutions to address dryland agricultural challenges and help improve the livelihood of smallholder farmers and rural communities.

Public-Private Partnerships (PPPs) are playing a growing role in agricultural innovation by combining public funding with private sector efficiency to accelerate the development and dissemination of market-driven innovations.

- A strong policy and regulatory environment is crucial, along with effective financing options and protection of intellectual property rights.
- Well-defined goals and guidelines are necessary for effective collaboration.
- Regular monitoring and evaluation are important to track progress and ensure accountability.
- Transparency
- Capacity building

MOA

A Memorandum of Association (MOA) is a legal document prepared in the formation and registration process of a limited liability company to define its relationship with shareholders. MOA is accessible to the public and describes the company's name, physical address of registered office, names of shareholders and the distribution of shares.

MOU

Defines a "general area of understanding" within both parties' authorities and no transfer of funds for services is anticipated. MOUs often state common goals and nothing more.

MOU may be used to outline the operation of a program so that it functions a certain way. For example, two agencies that have similar goals may agree to work together to solve a problem or support each other's activities by using an MOU. The MOU is nothing more than a formalized handshake.

Example

- A Memorandum of Understanding (MoU) was signed between the Indian Council of Agricultural Research (ICAR) and Dhanuka Agritech Limited with the aim of combining the strengths of both organizations to effectively transfer new technologies to farmers. Under this agreement, Dhanuka Agritech will collaborate with ICAR's central institutes, Agricultural Technology Application Research Institutes (ATARIs), and Krishi Vigyan Kendras (KVKs) to provide training to small-scale farmers on various aspects of agricultural production.
- a-IDEA of ICAR-NAARM signs MoA with 15 ICAR Institute 8th July, 2022, Hyderabad.

Scaling Up of Technologies

Definition

Scaling up refers to the intentional process of extending the reach and influence of innovations that have shown success in pilot or experimental settings. The goal is to ensure that these innovations benefit a larger population and contribute to long-term policy and program development.

As defined by Hartmann and Linn (2008), scaling up involves the expansion, adaptation, and continuation of effective policies, programs, and projects across different locations and over time, ultimately aiming to reach more people.

Four Key Components of Scaling Up

Outreach: This represents the extent or number of individuals or organizations that gain from the intervention.

Outcomes: These are the positive changes or impacts that the intervention is designed to achieve.

Sustainability: This component ensures that the positive effects of the intervention continue even after the project ends.

Equity: This emphasizes inclusive development, focusing on how the intervention creates opportunities for the most disadvantaged or marginalized groups in society.

Conceptual Model of Scaling Up

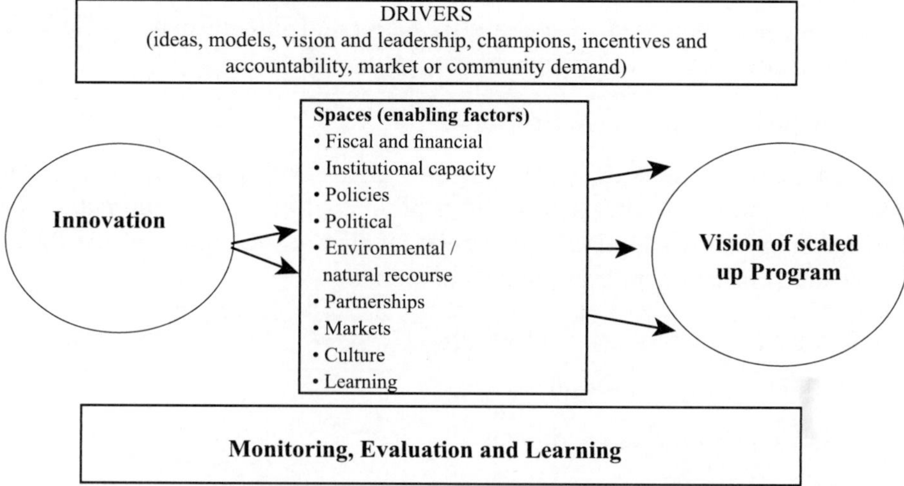

Pathways for Scaling Up

According to Linn (2012), a scaling-up pathway is a structured sequence of actions designed to move a successful innovation or pilot project from its trial phase to broader implementation. This pathway typically includes five key steps:

- Determining the type of scaling desired
- Promoting and advocating for the innovation
- Focusing on internal organizational processes
- Securing funding and mobilizing resources
- Conducting monitoring and evaluation to guide progress

Types of Scaling Up

Scaling up is accomplished by recognizing and promoting progress along different transformation pathways that lead to wider outcomes and long-term impact. There are several pathways to achieve scale, and they often complement one another. These approaches help highlight the various facets of scaling efforts.

Expansion or Replication (Horizontal Scaling or Scaling Out)

This involves extending the innovation to new geographical areas or introducing it to a broader or entirely new group of beneficiaries.

Policy, Political, Legal, or Institutional Scaling (Vertical Scaling)

This occurs when the innovation is officially adopted by government authorities at national or subnational levels and is integrated into formal development strategies and institutional frameworks.

Diversification (Functional Scaling)

This pathway focuses on enhancing an existing program by testing and integrating additional interventions. It is often applied once an innovation has reached substantial adoption and support, indicating its potential for further growth and enrichment through new components.

Challenges in Technology Scaling up

1. Initial investment is high
2. Scaling up process is complex
3. Market acceptance

What is Scalability?

Scalability – Concept and Measurement

As defined by Holcombe (2012), scalability refers to the likelihood that a specific innovation or change can be expanded, adapted, or replicated effectively. It reflects the confidence level that a proven innovation can achieve its full potential impact when taken to scale.

Steps in Measuring Scalability

Scalability is measured by evaluating both the innovation itself and the context in which it will be scaled. This includes:

1. Assessing the characteristics of the innovation and identifying essential enabling conditions
2. Reviewing how prepared key stakeholders and partners are for scale
3. Evaluating the presence of key conditions and calculating a scalability index
4. Developing an action plan to address any identified gaps or weaknesses

This process consists of **eight steps**:

- **Step 1:** Identify the critical conditions required to support scalability.
- **Step 2:** Examine and refine the questions used to assess each selected condition.
- **Step 3:** Assign scores to individual attributes related to each condition.
- **Step 4:** Calculate the effectiveness ratio using the formula:

Effectiveness Ratio = Actual Score/Maximum Possible Score

This ratio indicates how well each condition is currently fulfilled in terms of its impact on scalability.

- **Step 5:** Distribute points across the different sufficient conditions.
- **Step 6:** Compute the overall **scalability index**.
- **Step 7:** Compile a summary report based on comments and observations gathered during the assessment.
- **Step 8:** Evaluate the overall scalability of the innovation and determine the next steps.

Interpretation of the Scalability Index

- **Above 75**: **High Scalability** – Most necessary conditions are already present. Scaling up can be done with limited additional resources or effort.
- **Between 50 and 75**: **Moderate Scalability** – Several challenges may need to be addressed for successful scaling. Corrective measures should be taken before or during implementation.
- **Below 50**: **Low Scalability** – Major gaps exist. Significant work is required to create the necessary conditions before attempting to scale.

Technology Licensing

Intellectual Property and Technology Licensing – Paraphrased

Ideas, inventions, and other creative outputs are turned into private property and legally protected through the intellectual property (IP) system. **Licensing** plays a key role in this system by generating revenue, expanding the reach of a technology to a broader audience, and encouraging further innovation and commercialization.

- The **licensor** is the individual or organization that grants permission to use the IP.
- The **licensee** is the party that receives the rights to use the IP.

Technology, in this context, refers to the outcomes of scientific research and development, often taking the form of inventions.

Technology licensing allows the owner of such IP to permit another party to use it, while still retaining ownership. This is a major method of deriving value from intangible assets.

In a **technology licensing agreement**, the licensor provides IP rights to the licensee. These agreements usually include clauses related to royalties, performance standards, technical support, improvements, and termination conditions.

Six Key Concepts of Technology Licensing

1. Licensing happens only when one party possesses valuable intangible assets, known as intellectual property.
2. There are multiple forms of technology licenses, depending on the arrangement.
3. Licensing typically takes place within a broader business relationship, often accompanied by additional agreements.
4. The negotiation process involves both parties with differing interests, but who must find common ground.
5. Reaching an agreement requires resolving multiple, often complex, terms.
6. Licensing is related to, but not the same as, technology transfer.

Important Terms in Technology Licensing

- **Licensee**: The entity granted the right to use the technology.
- **Licensor**: The entity that owns the technology and grants the license.
- **Recitals**: A section outlining the background and purpose of the agreement.
- **Intellectual Property (IP)**: Includes patents, trademarks, copyrights, and trade secrets.
- **Royalties**: Payments made by the licensee to the licensor for the right to use the IP.

- **Territory**: The specific geographic area where the license is valid.
- **Term**: The length of time the license agreement is in effect.
- **Exclusivity**: The licensee's exclusive right to use the IP in a specific region or industry.
- **Non-Disclosure Agreement (NDA)**: A separate document to protect confidential information.
- **Warranty**: A promise that the technology meets agreed-upon specifications or standards.

Types of Licensing

- **Licensing-in**: When a company acquires technology from an external source—like a research lab, another company, or a university.
- **Licensing-out**: When a company grants another party the right to use its technology, usually for manufacturing, further development, or expansion.
- **Cross-licensing**: A mutual arrangement where two parties grant each other rights to use their respective technologies.

Exclusive License

- This type of license is granted to only one company.
- Since exclusive rights offer preferential access, the license fee or royalty is negotiated accordingly and finalized through mutual agreement between the licensor and licensee.

Non-Exclusive License

- This license can be granted to multiple firms, allowing broader manufacturing and commercialization.
- The licensing fee in this case is more flexible and can be adjusted based on terms and market conditions.

Technology Licensing Process

Technology licensing is coordinated by AgrInnovate India Limited (AgIn) in collaboration with the respective Institute Technology Management Unit (ITMU) or Zonal Technology Management Unit (ZTMU).

1. Disclosure of ICAR Technologies: The ITMU shares the key features of technologies available for commercialization with AgIn.
2. Techno-Commercial Assessment: Upon receiving the Technology Disclosure Form, AgIn forms a Techno-Commercial Assessment Committee to evaluate the technology's technical feasibility, market potential, necessary support, and to determine a suitable price.
3. Technology Evaluation and Standardization of Terms: A panel of experts reviews the technology, assessing its operational, economic, legal, and environmental suitability. Standard terms for licensing are developed in coordination with the concerned institute.
4. Business Development and Client Identification: AgIn undertakes promotional and business development activities to identify and engage with potential licensees.
5. Expression of Interest: Interested firms communicate their willingness to license the technology to AgIn.
6. Client Due Diligence: AgIn, along with ITMU/ZTMU, collects proposals from interested clients detailing how they plan to commercialize the technology. They also verify client credentials like company registration, GST status, and other compliance factors.
7. Agreement Finalization: Once both parties agree on the terms, a tripartite license agreement is executed among AgIn, the research institute, and the client.

Negotiating a Technology License Agreement

Negotiation in technology licensing involves creating a mutually acceptable agreement where the licensor permits, and the licensee obtains, the right to use a technology under defined terms.

The success of such negotiations depends on clearly understanding the value of the technology, along with the goals, expectations, and interests of both parties involved.

Negotiation process

Preparing

- Licensor and licensee would have identified each other as likely partners having the potential to complement, strengthen and fulfil each other's business aspirations.
- Identify what one wants to achieve from the discussions or what would be considered a successful outcome.
- A crucial step in the licensing process is to outline the main commercial elements that need to be addressed in the agreement. Each party licensor and licenseeshould clearly define their stance on these issues to guide the negotiation.

Discussion Phase

- The **licensor** highlights the strengths and potential benefits of the technology being offered.
- The **licensee** examines the provided information and related documents, typically under the terms of a confidentiality (non-disclosure) agreement.

Proposal Phase

- During this phase, both parties begin to explore the nature of their potential partnership and discuss core business terms.
- This includes asking critical questions, validating assumptions, identifying strategic goals, and setting clear limits and expectations.

Negotiation or Bargaining Phase

- In this stage, both sides engage in detailed discussions to align their respective offers and counteroffers.

- The objective is to reach a balanced agreement that fulfills the commercial interests of both parties.

Drafting Technology Licensing Documents

Creating the licensing agreement involves drafting a detailed and formal document that specifies all terms and conditions under which the technology is licensed. This includes clauses on rights granted, financial terms, performance obligations, confidentiality, duration, termination, and dispute resolution.

Terms of Trade and Agreement Process

Initiating Discussions

- Once a client shows interest in acquiring the technology, AgIn/ITMU/ZTMC will establish a committee to engage in discussions with the client. These discussions will be guided by a predefined framework, following standard terms of trade.
- These discussions will serve as the foundation for drafting the formal agreement.

Drafting and Legal Review of the Agreement

- A draft agreement will be created in alignment with the agreed terms.
- The draft will be circulated to both the respective institution and the client for their review and approval.
- After getting their approval, the institution or its nominee will arrange for legal vetting.
- The legal vetting will be carried out by an empanelled lawyer or law firm.

Agreement Signing

- The final agreement will be signed by the authorized representatives of all involved parties.
- The transaction will then be presented to the Agri Innovate board for formal acknowledgment.

Post-Agreement Support

- The concerned institute will handle the technology transfer and provide necessary handholding support as per the agreement's terms.

Completion Acknowledgment

- A review committee will be formed by AgIn/ITMU/ZTMC to evaluate the implementation after the agreement has been signed.

Managing the Technology Licensing Process

Strive for a Win-Win Outcome

- It's essential that all parties find the deal fair. Licensing agreements often lead to long- term relationships, so dissatisfaction can cause serious issues down the line and negate any prior gains.

Leverage Asymmetrical Trade-offs

- Focus on offering concessions that are inexpensive to you but highly valuable to the other party. For example, a garage may agree to repair a car at minimal cost to them but significant value to the customer. This type of trade leads to the most beneficial outcomes for both sides.

Set Negotiation Boundaries

- Define your maximum (best) and minimum (worst acceptable) positions on each issue before negotiations begin.
- However, these boundaries aren't rigid rules. If new information arises or gains are made in other areas, it may be reasonable to adjust your stance—even below the initially set minimum.

Aim High While Remaining Credible

- Start with strong, but realistic offers to keep your credibility intact. For example, lowballing an offer too far below market price can alienate the other party.
- Instead, propose a reasonable deal (e.g., a slight discount) and then seek additional benefits like extended warranty, free services, or upgrades—introduce other negotiable variables to enhance the final outcome without compromising the core value.

Technology Takers and Entrepreneurship

The Adoption Adaptation Strategy Matrix

The Adoption adaptation strategy matrix: maker, taker, tinker, and tailor

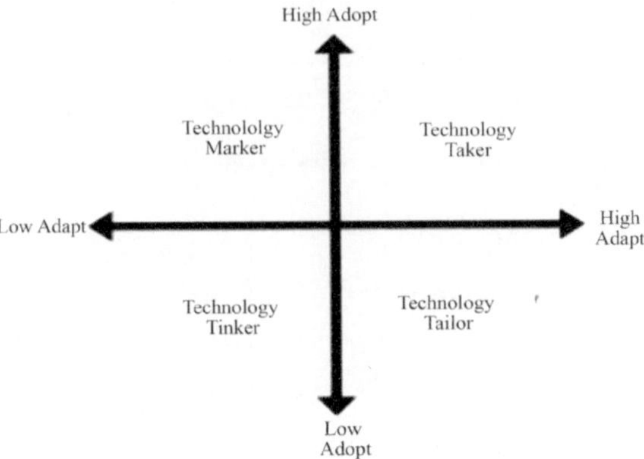

1. Technology Tailor

- Organizations following the **technology tailor strategy** adopt new technologies but insist on customizing them heavily to align with existing, often outdated, internal processes.
- They show little interest in changing behaviors or business models to match digital-era innovations.
- As technology evolves, tailors are forced into constant cycles of recustomization, making it difficult to keep up with new advancements and making upgrades increasingly complex and inefficient.
- Tailor organizations are deeply attached to their established processes and view these as untouchable.
- They frequently argue that they should determine which technology best fits their needs, rather than being expected to conform to the capabilities of the latest innovations.
- However, like an overly tailored suit, the resulting solutions are usually a poor fit in the digital age—they are customized, non-standard, and don't benefit from automatic updates.

2. Technology Tinker

- Tinkers avoid both adopting modern technologies and changing organizational behavior to fit new trends.
- This strategy reflects either a deliberate rejection of digital transformation

or an unwillingness to evolve, relying instead on outdated practices and technologies.

- These organizations may use some tools, but typically ones that are obsolete or no longer aligned with current digital standards.
- Over time, technology tinkers risk being **displaced** by competitors who embrace modern tools and approaches.

3. Technology Maker

- Technology makers are **innovators**—they create new technologies for others to use and adapt.
- Companies like **Uber, Apple, Google, and Facebook** are prime examples—they don't just use tech, they invent it and set the pace of digital change.
- Makers don't need to adjust their operations to fit others' innovations; instead, they define the standard.
- Some tailors mistakenly believe they are makers because they invest heavily in customizing existing technologies. However, customization does not equate to innovation or wide market applicability.
- True makers develop scalable, adaptable tools that others can implement.

Technology Taker Strategy – Guiding Principles

Principle 1: Start with a Technology-First Approach

- Organizations should **default to adopting existing digital-era technologies** as their primary strategy for both planning and execution.
- These modern tools are usually more efficient and effective than older, manually driven processes.

Principle 2: Prioritize Technology Over Process and People

- In this strategy, technology is seen as the **key driver of transformation**, rather than internal processes or workforce structures.
- By embracing tools created by tech makers, organizations can improve efficiency and avoid being overtaken by more agile competitors.
- If a company fails to adapt, others that do adopt new tech will gain the upper hand and displace slower movers.

Principle 3: Lead with Data-Driven Management

- Technology takers should focus on **analyzing data** generated by digital tools to improve strategic decision-making.
- Traditional performance indicators like profit/loss or sales pipelines are not enough in today's data-rich environment.
- By leveraging real-time, transactional data from business operations, organizations can track progress, test strategic hypotheses, and adjust their course dynamically.
- Proper investment in **data analytics** enables organizations to unlock the full potential of digital-era technologies.

Entrepreneurship

Entrepreneurship is the act of initiating and running a business or undertaking activities that generate economic value, with the primary aim of earning profit or delivering value.

- Innovation, ability to accept change and risk and the organization of resources are some of the factors involved in creating a sustainable enterprise.

Qualities of Entrepreneurship

- Self confidence
- Passionate and focused
- Risk taking ability
- Leadership ability
- Future oriented
- Creative

Types of Entrepreneurships

Traditional Entrepreneurship

- Traditional entrepreneurship focuses solely on generating profit and standardization. Instead of bringing a solution or an innovative idea to solve issues around us, commercial entrepreneurship operates based on the generation of profit without actually taking into consideration the sustainability of their actions.

Social Entrepreneurship

- This type of entrepreneurship focuses on producing product and services that resolve social needs and problems. Their only motto and goal is to work for society and not make any profits.

Green Entrepreneurship

- Green entrepreneurs develop environmentally sustainable products or services, promoting eco-friendly practices and reducing carbon footprints.

Imitative Entrepreneurship

Imitative entrepreneurship refers to the practice of replicating existing products or services already available in the market, often through franchise agreements. This type of entrepreneurship plays a crucial role in the global diffusion of technology. It involves adopting proven technologies in different countries, sometimes with slight adaptations to suit local conditions and needs.

Agripreneurship

An agripreneur is someone who launches, manages, and operates a business venture within the agricultural sector.

In a broader sense, agri-entrepreneurship or agripreneurship focuses on adding value to agricultural resources, typically engaging rural populations and utilizing their labor and skills.

Agricultural entrepreneurship shares many characteristics with other types of entrepreneurship, such as commercial ventures, eco-friendly businesses, social enterprises, and sustainable models.

There are four primary service areas in agripreneurship:

1. Providing expert advice on crop production
2. Supplying agricultural inputs
3. Establishing market connections
4. Facilitating access to credit

Agripreneurship spans a variety of agricultural activities, including dairy farming, sericulture, goat and rabbit rearing, floriculture, fisheries, shrimp farming, sheep rearing, vegetable cultivation, nursery operations, and even farm forestry.

Agripreneurs also establish and run enterprises such as:

- Agro-processing units (e.g., rice mills, pulse mills)
- Agro-product manufacturing units (e.g., sugar or bakery production)
- Agro-input production facilities (e.g., fertilizer and food processing plants)
- Agro-service centers (e.g., repair hubs for agricultural machinery)
- Other ventures including beekeeping (apiaries), seed and feed processing units, mushroom cultivation, compost production, organic fruit and vegetable stores, and even the cultivation of biofuel crops like jatropha.

Modern agripreneurship models increasingly connect farmers with both rural and urban markets through decentralized approaches. These models aim to encourage youth involvement in entrepreneurship and help stimulate local and regional economies.

Startups

A start-up is a venture that is initiated by its founders around an idea or a problem with a potential for significant business opportunity and impact. The term start-up refers to a company in the first stages of operations.

Lifecycle of startups

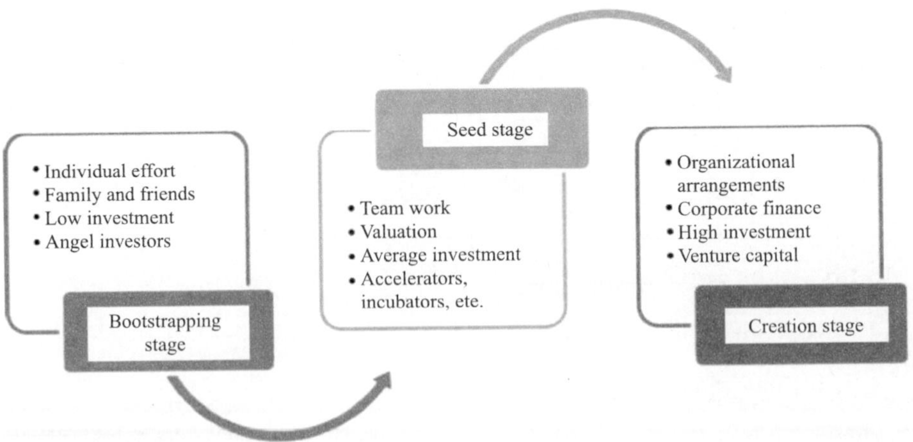

(i) Bootstrapping stage

At this initial phase, the entrepreneur takes the initiative to transform an idea into a viable and profitable business. **Bootstrapping** refers to the use of highly resourceful and innovative strategies to utilize resources without taking on debt.

The main goal of this stage is to lay the foundation for future growth by proving the product's feasibility, managing cash flow effectively, building and managing a team, and gaining initial customer validation.

Angel investors are often more inclined to provide funding during this phase.

The phrase, *"Bootstrapping is a way of life in entrepreneurial companies,"* highlights the reason many startup theories stem from broader entrepreneurship frameworks.

(ii) Seed stage

This stage involves forming a team, developing a prototype, entering the market, assessing the value of the startup, and pursuing support systems like **incubators** and **accelerators**. Seed funding is the initial investment used to develop a product or service. Entrepreneurs typically seek assistance from support organizations such as **accelerators, incubators, business development centers**, and **hatcheries** to fast-track growth. A large number of startups do not survive this phase, often due to the lack of access to these support systems. At best, many may end up becoming low-profit ventures with limited chances of success. Valuation of the startup is generally conducted toward the end of the seed stage.

(iii) Creation stage

In this phase, the company begins commercial operations by selling products, entering the market, and hiring its first employees.

This further supports the idea that most startup theories are grounded in **entrepreneurship**, rather than **management or organizational theory**.

By the end of this stage, the business entity is officially formed, and **corporate financing** becomes a key method for raising capital. **Venture capitalists** often play a crucial role in supporting ventures at this point by providing necessary funding.

Small Businesses

To address the specific needs of micro, small, and medium enterprises (MSMEs), there was a need to establish a unified legal structure. The **Micro, Small and Medium Enterprises Development (MSMED) Act, 2006** was introduced to address critical areas such as defining MSMEs, improving access to credit, supporting marketing efforts, and facilitating technology upgrades.

Characteristics of Small Business Entrepreneurship

- Entrepreneurs focus on a single product, technology, market or localism while establishing small business entrepreneurship.
- Small business entrepreneurship usually has no expansion plans in the initial stages of establishment.
- The primary objective of small business entrepreneurship is to generate profits from the early days of establishment.

Enterprises are classified into two major categories viz., manufacturing and services

A. Manufacturing

Classification of Enterprises Based on Investment

A. Manufacturing/Production Sector

For businesses involved in manufacturing or producing goods within any industry, enterprises are classified as follows:

- **Micro Enterprise**: Investment in plant and machinery does not exceed ₹ 25 lakh.
- **Small Enterprise**: Investment is more than ₹ 25 lakh but does not exceed ₹ 5 crore.
- **Medium Enterprise**: Investment exceeds ₹ 5 crore but is capped at ₹ 10 crore.

B. Service Sector

For businesses offering or delivering services, the categorization is:

- **Micro Enterprise**: Investment in equipment is up to ₹ 10 lakh.
- **Small Enterprise**: Investment exceeds ₹ 10 lakh but does not surpass ₹ 2 crore.
- **Medium Enterprise**: Investment is over ₹ 2 crore and up to ₹ 5 crore.

Producer Organisation (PO)

A **Producer Organisation (PO)** is a legally recognized entity created by primary producers such as farmers, weavers, milk producers, artisans, and fishermen. POs can be structured as producer companies, cooperative societies, or other legal forms that allow for shared profits or benefits among their members. In certain models, even institutions representing producers can be part of the PO.

The core goal of a PO is to improve the income levels of its members by organizing small- scale producers who may not independently have sufficient input or output volume to benefit from economies of scale. In agricultural markets, long and opaque supply chains often result in producers receiving only a small fraction of the final consumer price. POs aim to address this imbalance.

Key Features of a PO

- Formed by producers for both agricultural and non-agricultural activities.
- Registered as a legal entity.
- Members are shareholders of the organization.
- Engages in commercial activities related to the members' produce.
- Operates in the interest of its members.
- Shares a portion of profits with its members.
- Reinvests the remaining surplus for business development.

Major Functions of a PO

- Procuring agricultural inputs.
- Sharing market intelligence.
- Promoting technologies and innovations.
- Assisting in obtaining input financing.
- Collecting and storing produce.
- Conducting primary processing like cleaning, drying, and grading.
- Creating branding, packaging, and labeling.
- Maintaining quality standards.
- Selling to bulk buyers or institutions.
- Engaging with commodity exchanges.
- Participating in export markets.

Self-Help Groups (SHGs)

Self-Help Groups (SHGs) are informal, community-based groups formed voluntarily by individuals seeking to improve their livelihoods. They are peer-regulated and self-managed, typically consisting of people with similar socio-economic backgrounds and shared objectives.

Rural communities often face challenges such as poverty, illiteracy, lack of access to formal credit, and limited skill development. SHGs enable collective efforts to overcome these barriers. SHGs are built on the principle of "self-help," encouraging entrepreneurship and strategies to reduce poverty.

Advantages of SHGs

- **Social Unity**: They foster collective action to address issues like dowry, alcoholism, and other social evils.
- **Gender Equality**: SHGs empower women by nurturing leadership skills and encouraging participation in local governance, such as Gram Sabha and elections.
- **Women's Empowerment**: Evidence shows SHGs significantly improve women's socio-economic status and self-confidence, both in the household and in society.
- **Advocacy and Participation**: SHGs serve as platforms for marginalized groups to voice concerns on public issues like sanitation, healthcare, and education, influencing local policy decisions.
- **Financial Inclusion**: SHGs promote access to formal banking channels. Through the **SHG-Bank Linkage Programme** initiated by **NABARD**, group members gain easier access to credit, reducing dependency on informal lenders. Banks are encouraged to support SHGs under **Priority Sector Lending** due to reduced risk and reliable returns.

Clustures

A business cluster is usually "a geographic concentration of interconnected companies, specialised suppliers, service providers, and associated institutions in a particular field that is present in a nation or region."

(i) **Opportunities:** Access to a larger skilled labour pool- being part of a cluster grouping can bring easier access to employees, training programs, institutions and services.

(ii) **Competitive advantage:** local companies gathered in clusters have an advantage over international companies because they have better access to the market, know the market well and thus can have a closer relationship with their customers. The close proximity of the local companies to their customers brings also benefits in terms of a quick delivery and savings on travel costs.

(iii) **Innovation:** As clusters provide the opportunity for knowledge sharing, it enables companies to better understand customers, get to know their needs and habits and thus create more personable relationships with them. This strong connection between the company and its customers enables it to respond swiftly to market changes, identify emerging needs, and quickly adapt or innovate products and services to align with customer expectations.

(iv) **Less competition:** when competitors from the closely related industries decide to join forces and collaborate in the new business unit, they simultaneously eliminate the competition that can thus lead to higher profits and net incomes. Apart from wages, well-developed Clusters in Business many companies are trying to gain a competitive advantage in any possible way to make their business profitable, especially when they see that the potential for growth is increasing.

(v) **Regional development:** as companies in business clusters start to grow, they need to hire more people to sustain their position in the market. Thus, cluster development affects regional development as it creates new job vacancies and higher tax revenues for the area.

Policy Support for Technology Commercialisation and Entrepreneurship Development: Industrial Policy

It encompasses the guidelines, laws, principles, policies, and procedures established by the government to oversee and manage industrial enterprises within the country. It prescribes the respective role of public, private, joint, cooperative, large, medium and small-scale sectors for the development of industries. It includes fiscal policy, monetary policy, tariff policy as well as policy for labour. It also shows government attitude towards foreign investment.

Main Objectives

- To maintain a sustained growth in productivity.
- To enhance employment opportunities.
- To prevent concentration of economic power.
- Optimum utilisation of human resource.

- To achieve competitiveness in international area.
- To transform India into global power.

Major Industrial Policies

- Industrial Policy Resolution 1948
- Industrial Policy Resolution 1956
- Industrial Policy Resolution 1977
- Industrial Policy Resolution 1980
- New Industrial Policy 1991
- Industrial Policy Resolution 2000

Policy Initiatives for Technology Commercialization in India

1. Strengthening Industry–Academia Collaboration

- There is a growing understanding among academic and research institutions in India about the need to work closely with industries, despite cultural and operational differences.
- However, the absence of a well-defined legal structure for technology transfer and ambiguous intellectual property (IP) practices has limited proactive collaboration. Most existing partnerships are formalized through Memoranda of Understanding (MoUs), often favoring private sector interests in terms of knowledge ownership.
- Over recent years, increased focus has been placed on IP management in public research institutions. This has led to the adoption of institutional IP policies and frameworks by major national research bodies like CSIR, ICMR, ICAR, as well as leading academic institutions such as IIT Bombay, IIT Delhi, and IISc Bengaluru.

Example Initiative: National Agricultural Innovation Project (NAIP)

- The Government of India, with World Bank support, launched NAIP in 2006 for a six- year period (ending in 2012), implemented by ICAR.
- It builds upon the earlier National Agricultural Technology Project (NATP) and emphasizes collaborative development and diffusion of agricultural innovations.
- NAIP established consortia composed of research institutions, industry, and development organizations working along the agricultural value chain—from input provision to crop harvesting.

2. Promoting Research in Public Institutions
    - India has recognized the need for legislation supporting the commercialization of publicly funded science.
    - A policy similar to the U.S. Bayh-Dole Act is being developed in the form of the Public Funded R&D (Protection, Utilization and Regulation of Intellectual Property) Bill, which has received parliamentary approval and is in the process of formalization.
3. Enhancing Knowledge Transfer Mechanisms
    - In 2005, key public research bodies such as CSIR, ICAR, ICMR, and DBT created a platform for critical discussions on technology transfer.
    - The Society for Technology Management (STEM) has expanded significantly, gaining over 100 new members in the last three years.
    - STEM includes not only Indian institutions but also members from research organizations in Southeast Asia and the Middle East, along with major life sciences firms from the U.S., Europe, and India.
4. Establishing Technology Business Incubators (TBIs)
    - The Ministry of Science and Technology, through the National Science and Technology Entrepreneurship Development Board (NSTEDB), is spearheading initiatives to foster entrepreneurship.
    - TBIs and Technology Parks are being developed within reputed technology institutions, including IITs (Delhi, Chennai, Mumbai, Kanpur), Delhi University, and IT-BHU (Varanasi).
    - A 2010 Task Force from the Prime Minister's Office recommended the establishment of around 100 TBIs in top national science and technology institutions.

National Policy on Skill Development and Entrepreneurship (2015)

The National Policy on Skill Development and Entrepreneurship, 2015 was designed to address the challenge of providing large-scale skill training quickly and with high quality. It aims to unify various skill initiatives under a common framework, aligning them with market needs.

The policy outlines clear objectives, expected outcomes, and identifies key institutions that will facilitate implementation. It also aims to integrate skill development with employability and productivity improvements.

Key Government Initiatives Under this Policy Include

1. Pradhan Mantri Kaushal Vikas Yojana (PMKVY): Offers short-term vocational training for youth.
2. Standardization through Common Norms: Ensures consistent implementation across skill development schemes by various ministries.
3. Establishment of Sector Skill Councils (SSCs): Industry-led bodies responsible for developing National Occupational Standards (NOSs).
4. SANKALP (Skills Acquisition and Knowledge Awareness for Livelihood Promotion): A World Bank-backed project focused on strengthening skill institutions at both national and state levels.
5. STRIVE (Skills Strengthening for Industrial Value Enhancement): Another World Bank-supported program aimed at improving the relevance and quality of training provided by ITIs and apprenticeships.
6. Launch of the Skill India Portal: A centralized platform offering information about trainees, trainers, and training institutions.
7. Mobilization and Outreach Programs: Activities such as Kaushal Melas (skill fairs), Rozgar Melas (job fairs), and career counseling schemes help make vocational training more aspirational and accessible.

National Innovation and Startup Policy (NISP)

The 'National Innovation and Start-up Policy 2019 for students and faculty in Higher Education Institution HEIs was launched by Hon'ble Minister of Education, Shri Ramesh Pokhriyal 'Nishank' on 11th September 2019 at All India Council for Technical Education AICTE, New Delhi.

Policy Objectives

- The National Innovation and Start-up Policy 2019 intends to guide HEIs to promote student driven innovations & start-ups and to engage the students and faculty in innovation and start up activities in campus.
- The policy aims at enabling HEIs to build, streamline and strengthen the innovation and entrepreneurial ecosystem in campus and will be instrumental in leveraging the potential of science, student's creative problem solving and entrepreneurial mind-set, and promoting a strong intra and inter-institutional partnerships with ecosystem enablers and different stakeholders at regional, national and international level.

The Government of India has launched various initiatives and implemented policy measures to promote innovation and entrepreneurship across the country. Addressing the challenge of job creation remains a top priority. Given its distinct demographic advantage, India holds vast potential to drive innovation, cultivate entrepreneurship, and generate employment that benefits both the nation and the global community.

In recent years, the government has introduced a broad range of programs and opportunities aimed at fostering innovation across multiple sectors. These efforts involve collaboration with academic institutions, industries, investors, entrepreneurs of all scales, non-governmental organizations, and even the most marginalized communities.

Make in India

Make in India is an initiative introduced by the Government of India in 2014 under the leadership of Prime Minister Narendra Modi. Its primary goal is to promote and support businesses in developing, producing, and assembling products within India, while also attracting targeted investments in the country's manufacturing sector. The policy approach was to create a conducive environment for investments, develop a modern and efficient infrastructure, and open up new sectors for foreign capital. The initiative targeted 25 economic sectors for job creation and skill enhancement, and aimed "to transform India into a global design and manufacturing export hub.

Make in India" had three Stated Objectives

1. To increase the manufacturing sector's growth rate to 12-14% per annum;
2. To create 100 million additional manufacturing jobs in the economy by 2022;
3. To ensure that the manufacturing sector's contribution to GDP is increased to 25% by 2022 (later revised to 2025).

Stand up India

Stand up India Scheme was launched on 5th April 2016 to promote entrepreneurship at grassroot level focusing on economic empowerment and job creation. The Stand-Up India scheme aims to tackle the difficulties encountered by SC, ST, and women entrepreneurs in starting businesses, securing loans, and accessing necessary support for achieving success in their ventures. The scheme therefore endeavours to create an ecosystem which facilitates and continues to provide a supportive environment to the target segments in doing business. The scheme aims to encourage all bank branches in extending loans to borrowers from SC, ST and women in setting up their own greenfield enterprise.

Startup India

Launched in January, 2016. Startup India Initiative has rolled out several programs with the objective of supporting entrepreneurs, building a robust startup ecosystem and transforming India into a country of job creators instead of job seekers.

These initiatives are overseen by a specialized Startup India Team under the Department for Promotion of Industry and Internal Trade (DPIIT). The program offers a free, comprehensive four-week online learning course and has established research parks, incubators, and startup hubs nationwide by fostering strong collaborations between academic institutions and industry organizations.

Significantly, a **Fund of Funds** has also been established to enhance funding opportunities for startups. Beneficiaries of Startup India include entrepreneurs, startups, small and medium enterprises, investors, and skilled workers, all receiving support through funding, mentorship, and simplified regulations.

Atal Innovation Mission

The Atal Innovation Mission (AIM), launched in 2018, is a flagship initiative of the Government of India aimed at fostering and nurturing a culture of innovation and entrepreneurship throughout the country. AIM's objective is to develop new programmes and policies for fostering innovation in different sectors of the economy, provide platforms and collaboration opportunities for different stakeholders, and create an umbrella structure to oversee the innovation & entrepreneurship ecosystem of the country.

Digital India

Launched on 1 July 2015, by Indian Prime Minister Narendra Modi to ensure that the Government's services are made available to citizens electronically through improved online infrastructure and by increasing Internet connectivity or making the country digitally empowered in the field of technology. The initiative includes plans to connect rural areas with high-speed internet networks. It consists of three core components: the development of secure and stable digital infrastructure, delivering government services digitally, and universal digital literacy.

RKVY

The "Innovation and Agri-Entrepreneurship Development" program was launched in 2018 as part of the Rashtriya Krishi Vikas Yojana (RKVY). The aim of the program is to promote innovation and agricultural entrepreneurship by providing both financial and technical support to startups. To facilitate this, 24 RKVY Agribusiness Incubators (R-ABIs) along with five Knowledge Partners (KPs) have been designated to offer training and incubation services for entrepreneurs and to oversee the program's implementation. Under this initiative, startups and entrepreneurs in the agriculture and allied sectors are eligible to receive funding of up to INR5 lakh at the idea or pre-seed stage, and up to INR 25 lakh at the seed stage.

Under the initiative, five key personnel (KPs) and twenty-four research and business incubators

(R-ABIs) are responsible for providing training and incubation to startups.

The Government of India conducts various national-level programs to establish a platform that supports the growth of agri-startups by connecting them with diverse stakeholders. These programs include agri-startup conclaves, agri-fairs and exhibitions, webinars, and workshops. Beneficiaries include farmers, agricultural entrepreneurs, startups, and research institutions, all receiving support to enhance agricultural innovation and productivity.

Institutions Supporting Entrepreneurship in India

The Indian government has established numerous institutions and training centres aimed at fostering entrepreneurship by enhancing the knowledge, skills, and mindset of aspiring entrepreneurs. Both central and state governments have developed specialized agencies to support entrepreneurial growth, including:

- Small Industries Service Institutes (SISI)
- Small Industries Development Organisation (SIDO)
- National Small Industries Corporation (NSIC)
- Small Industries Extension Training Institute
- Entrepreneurship Development Institute of India (EDII), Ahmedabad, Gujarat (Established in 1983)
- Institute for Rural Management and Administration
- National Institute for Entrepreneurship and Small Business Development (NIESBUD)

- National Alliance of Young Entrepreneurs (NAYA)
- Ministry of Micro, Small and Medium Enterprises (MSME)
- Maharashtra Centre for Entrepreneurship Development (MCED)
- West Bengal Industrial Development Corporation (WBIDC)
- Technology Development and Adaptation Centre (TDAC)
- Khadi and Village Industries Commission (KVIC)

These organizations play a vital role in nurturing entrepreneurial talent through training, financial support, and capacity-building initiatives across various sectors.

4

Technology Incubation

Basics of Technology Incubation

Technology incubation is a fascinating process that fosters the growth of innovative startups and businesses with high-potential technology. It provides a supportive environment and resources for early-stage companies to develop their ideas, validate their business models, and navigate the challenges of commercialization.

Here's a breakdown of the key aspects of technology incubation: What happens in a technology incubation?

Pre-incubation: Startups with promising ideas can receive initial guidance and support to refine their concepts and develop a business plan.

Incubation: Selected startups gain access to shared office space, mentorship from experienced professionals, networking opportunities, and access to funding sources.

Post-incubation: Graduated startups continue to receive support in areas like marketing, legal advice, and investor relations as they scale their businesses.

Benefits of Technology Incubation

Reduced risks: Startups can test their ideas and validate their market potential before investing significant resources.

Access to resources: Incubators provide vital resources like office space, equipment, and software, reducing the financial burden on startups.

Expert guidance: Mentorship from experienced entrepreneurs and industry professionals helps startups navigate challenges and make informed decisions.

Networking opportunities: Incubators connect startups with potential investors, partners, and customers, facilitating collaboration and growth.

Increased chances of success: Startups participating in incubation programs have a higher success rate than those operating independently.

Technology Incubation: Nurturing Innovation for Growth

Technology incubation refers to the process of supporting and fostering the growth of early- stage startups and innovative ventures by providing them with resources, infrastructure, and mentorship. It acts as a nurturing environment where fledgling companies can develop their ideas, refine their business models, and navigate the challenges of launching and scaling their operations.

Key Features of Technology Incubation

Shared Workspace: Incubators often provide shared office space, co-working facilities, and laboratories to reduce initial costs and foster collaboration among startups.

Financial Assistance: Some incubators offer grants, seed funding, or access to investors to help startups meet their financial needs.

Mentorship and Training: Experienced professionals provide guidance, advice, and training on various aspects of business management, marketing, and technology development.

Networking Opportunities: Incubators connect startups with potential investors, partners, and mentors, expanding their network and access to resources.

Business Support Services: Incubators may offer legal, accounting, marketing, and other support services to help startups navigate operational hurdles.

Types of Technology Incubators

University-based Incubators: Leverage university resources and expertise to support student entrepreneurs and faculty startups.

Sector-specific Incubators: Focus on specific industries or technologies, such as healthcare, biotechnology, or cleantech.

Private Incubators: Operated by venture capitalists or private investors to identify and nurture promising startups with potential for high returns.

Non-profit Incubators: Focus on social impact and supporting startups with the potential to address social or environmental challenges.

Stakeholder-Oriented Incubation Process

A Framework Understanding Stakeholder Orientation

A stakeholder-oriented incubation process recognizes that the success of a new venture is not solely dependent on the entrepreneur or the incubator. It involves a collaborative effort among various stakeholders, including:

- Entrepreneurs: The individuals or teams with innovative ideas.
- Incubator: The organization providing resources and support.
- Investors: Those providing financial backing.
- Mentors: Experienced individuals offering guidance.
- Customers: Potential users of the product or service.
- Partners: Collaborators, suppliers, and distributors.
- Government: Regulatory bodies and policymakers.
- Community: Local residents and organizations.

Key Steps in a Stakeholder-Oriented Incubation Process

1. Stakeholder Identification and Mapping

1. Identify all relevant stakeholders.
2. Map their relationships and interests.
3. Assess their potential impact on the venture.

2. Stakeholder Engagement and Communication

1. Develop a communication plan to engage stakeholders.
2. Use appropriate channels (e.g., meetings, surveys, social media).
3. Listen actively to their needs, concerns, and expectations.

3. Value Proposition Alignment

1. Ensure the venture's value proposition aligns with the needs and interests of key stakeholders.
2. Make necessary adjustments to the business model or product development.

4. Shared Goal Setting

1. Involve stakeholders in setting clear, measurable, achievable, relevant, and time-bound (SMART) goals.
2. Create a shared vision for the venture's success.

5. Resource Allocation and Collaboration

1. Allocate resources (e.g., funding, mentorship, infrastructure) based on stakeholder needs and priorities.
2. Foster collaboration among stakeholders to leverage their expertise and networks.

6. Continuous Monitoring and Evaluation:

1. Track progress towards goals and measure stakeholder satisfaction.
2. Make adjustments as needed to ensure the venture's success.

Benefits of a Stakeholder-Oriented Incubation Process

- **Increased Success Rate:** By addressing the needs and concerns of all stakeholders, the venture's chances of success are significantly improved.
- **Stronger Relationships:** Building positive relationships with stakeholders fosters trust, cooperation, and support.
- **Enhanced Innovation:** A diverse group of stakeholders can contribute valuable insights and ideas, leading to more innovative solutions.
- **Reduced Risk:** By anticipating and addressing potential challenges, the venture's risk profile is lowered.
- **Positive Impact:** A stakeholder-oriented approach can create a positive impact on the community and society as a whole.

Types of incubators

1. Technology business incubator
2. Agri business incubator
3. Livelihood business incubator
4. Village incubator

1. Technology Business Incubators (TBIs)

Technology Business Incubators (TBIs) are specialized organizations designed to nurture and support technology-based startups. They provide a supportive environment, resources, and mentorship to help these young companies grow and succeed.

Key Features of TBIs

- **Focus on Technology:** TBIs primarily cater to startups in the technology sector, including software development, hardware manufacturing, biotechnology, and information technology.
- **Infrastructure and Resources:** They offer access to state-of-the-art technology infrastructure, research facilities, and equipment.
- **Mentorship and Guidance:** Experienced entrepreneurs, industry experts, and mentors provide guidance and support to startups.
- **Networking Opportunities:** TBIs facilitate networking with investors, potential customers, and other industry professionals.
- **Education and Training:** They often provide workshops and training programs to enhance the business skills of entrepreneurs.
- **Funding:** Some TBIs may offer seed funding or help connect startups with investors.
- **Physical Space:** They typically provide office space or co-working facilities for startups.

Examples of Renowned TBIs

- **Y Combinator:** One of the most well-known TBIs in the world, known for its highly competitive application process and successful alumni.
- **Techstars:** A global network of TBIs that offers mentorship, funding, and resources to startups.
- **500 Startups:** An early-stage venture capital firm and incubator that invests in startups around the world.
- **Plug and Play:** A global innovation platform that connects startups with corporate partners and investors.

2. Agricultural Business Incubators (ABIs)

Agricultural Business Incubators (ABIs) are specialized organizations that support startups in the agriculture and food industry. They provide a nurturing environment, resources, and mentorship to help these businesses grow and succeed.

Key Features of ABIs

- **Focus on Agriculture:** ABIs cater to startups involved in various aspects of agriculture, including farming, food processing, agricultural technology, and sustainable agriculture.
- **Research and Development:** They often provide access to research facilities, agricultural experts, and resources for product development.
- **Market Access:** ABIs help startups connect with potential customers, distributors, and retailers.
- **Regulatory Compliance:** They offer guidance on compliance with agricultural regulations, certifications, and standards.
- **Networking Opportunities:** ABIs facilitate networking with investors, industry professionals, and other startups.
- **Education and Training:** They provide workshops and training programs on topics such as agribusiness management, sustainable farming practices, and food safety.
- **Physical Space:** ABIs may provide laboratory space, greenhouses, or other facilities for agricultural research and development.

Examples of Renowned ABIs

- **The Hatchery:** A food-focused incubator based in California that supports startups in the food industry.
- **The Food Foundry:** An incubator that focuses on sustainable food innovation and technology.
- **AgLaunch:** A nonprofit organization that supports agricultural startups in the southern United States.
- **AgTech Innovation Lab:** An incubator that focuses on agricultural technology and innovation.

Livelihood Incubation

Livelihood incubation takes a unique approach to fostering sustainable development by empowering communities, particularly in rural and underserved areas, to build and improve their livelihoods. It goes beyond the traditional focus on business incubation and instead prioritizes local knowledge, resource utilization, and community participation.

Key Features of Livelihood Incubation

Community-Driven: Local communities actively participate in identifying needs, choosing potential livelihood ventures, and developing strategies for sustainable livelihood generation.

Focus on Local Resources: Utilizing locally available resources like agricultural products, traditional skills, and cultural heritage to create sustainable livelihoods.

Skill Development and Capacity Building: Providing training and resources to equip communities with the necessary skills and knowledge to manage and grow their chosen livelihood initiatives.

Micro-Enterprise Promotion: Encouraging the establishment and growth of micro-enterprises that cater to local needs and markets.

Sustainability Focus: Emphasizing environmentally friendly practices and resource management for long-term sustainability of livelihoods.

Benefits of Livelihood Incubation

Poverty Reduction: Provides communities with an alternative path to income generation and poverty reduction.

Empowerment and Participation: Increases community ownership and participation in local development initiatives.

Entrepreneurship and Skill Development: Fosters entrepreneurial spirit and equips communities with valuable skills.

Resource Utilization and Sustainability: Promotes sustainable utilization of local resources and environmental protection.

Diversity and Resilience: Strengthens local economies and builds resilience against external shocks.

Examples of Livelihood Incubation

Honeybee Network in India focuses on promoting traditional knowledge and sustainable agriculture practices through community-managed enterprises.

SEWA (Self Employed Women's Association) in India empowers women through skill development and micro-enterprise creation in various sectors.

The REDD+ program supports efforts to reduce deforestation and emissions from deforestation and forest degradation, creating sustainable income opportunities for forest communities.

3. Village incubators

Village incubators are specialized programs designed to support entrepreneurship and economic development in rural areas. They provide a range of resources, training, and mentorship to help individuals and communities start and grow businesses. Village incubators play a vital role in bridging the gap between rural and urban areas by providing opportunities for entrepreneurship and economic development in underserved regions. By empowering individuals and communities, they can contribute to a more equitable and sustainable future.

Key Characteristics of Village incubators

1. **Community-Focused:** Village incubators are often rooted in local communities and prioritize the needs and aspirations of rural residents. They aim to foster a sense of ownership and empowerment among the local population.
2. **Holistic Approach:** They address a wide range of challenges faced by rural entrepreneurs, including access to finance, markets, technology, and skilled labor. This comprehensive approach ensures that startups have the necessary support to succeed.
3. **Sustainable Development:** Village incubators promote economic growth that is both sustainable and beneficial to the environment.
4. **Local Expertise:** They leverage local knowledge and resources to develop tailored solutions for rural entrepreneurs. By understanding the unique challenges and opportunities of rural areas, village incubators can provide more effective support.

Services Offered by Village Incubators

- **Business Planning:** Providing guidance on business plan development, market research, and financial analysis.
- **Skill Development:** Offering training in entrepreneurship, management, and technical skills.
- **Mentorship:** Connecting entrepreneurs with experienced mentors who can provide advice and support.

- **Networking Opportunities:** Facilitating connections with investors, customers, and other stakeholders.
- **Access to Finance:** Helping entrepreneurs secure loans, grants, or other forms of financing.
- **Technology Support:** Providing access to technology and training on its use.

Technology Incubation in India

The incubation process within a technology incubator typically follows a distinct series of phases, designed to support and guide startups from their initial stage to a point where they can successfully operate on their own.

Process/ Stages of Incubation at Technology Business Incubation (TBI)

The incubations process comprised of;

1. Pre-incubation Process
2. Incubation Process
3. Post –incubation or Accelerator process

Pre-Incubation Process: Selection and Inclusion Criteria for Talent in TBI Programs

The applicants to Technology Business Incubators (TBIs) come from varied backgrounds — ranging from recent graduates to startups less than a year old. Therefore, the selection approach must be inclusive and flexible, yet focused on fostering a sustainable knowledge-driven ecosystem. This applies across both community-based incubators (usually offering free or low- cost services) and commercial incubators (which often receive equity stakes in return for their support).

The typical process followed includes:

1. Application Submission by Talent

- Interested individuals or groups must submit a formal application, often through a standardized form designed by the incubator.
- Required documents may include basic details (names, team members), a business plan, registration and financial records (if available), recommendation letters, and other due diligence materials.

2. Initial Screening and Compliance Check

- The incubator conducts preliminary evaluations, including background and compliance checks, before forwarding the applications to the selection committee.

3. Evaluation by Selection Panel

- A designated panel of experts reviews the applications and identifies a shortlist of potential candidates.
- These selected applicants then proceed to the interview stage for further assessment.

4. Interview Round with Experts

- Shortlisted applicants are interviewed to assess their potential, suitability, and alignment with the goals and framework of the TBI.
- This step helps gauge whether the candidates are a good fit for the program.

5. Pre-Selection Notification

- The selected individuals or teams are informed about the terms of engagement— including roles, benefits, and responsibilities.
- This gives them time to evaluate the opportunity, compare other offers, and prepare for the next stage involving financial and operational negotiations.

6. **Negotiating Incubation Terms**

- While most incubators have a standard agreement template, specific terms (like financial contributions, equity shares, and incubation milestones) are usually customized through mutual negotiation.

7. **Agreement Signing and Onboarding:**

- Once the agreement is finalized, onboarding begins.
- This includes introductions to the incubator's staff, facilities, and operational framework, allowing the new talent to integrate smoothly and begin their journey toward accelerated growth within the set timeframe of their incubation contract.

Idea validation: This phase involves assessing the feasibility and market potential of the startup's idea. Potential startups might participate in workshops, pitch their ideas to mentors, and receive feedback to refine their concept.

Business plan development: Startups develop a comprehensive business plan outlining their target market, financial projections, and marketing strategies. Mentors and advisors can provide guidance during this process.

Team formation: The right team with complementary skills and experience is assembled to drive the venture forward. Incubators may help connect startups with potential team members.

2. Incubation process: Services provided by Incubators Location-Based Physical Services

Technology Business Incubators (TBIs) typically offer modern and well-equipped physical infrastructure, usually situated in a dedicated and isolated area. These facilities generally include:

- Contemporary office spaces.
- Communication tools like telephone, internet, and fax.
- IT resources such as desktops, laptops, and printers.
- Specialized tools or equipment tailored to the specific domain of the incubated project (provided as needed).
- Access to the host institution's library and information resources.
- Training and conference amenities that support collaboration, including domestic and international partnerships for both incubator staff and incubated teams.

Business-Oriented Services

For a TBI to be effective, it must offer a consistent set of business support services. These usually include:

1. Marketing support and market research.
2. Business development training and strategic planning assistance.
3. Arranging managerial and technical guidance.
4. Help in acquiring necessary government or regulatory approvals.
5. Sharing of information on new product ideas and technologies.
6. Assistance with financial planning and securing funding.
7. Support for legal and intellectual property rights (IPR) matters.
8. Access to the host institution's facilities at minimal cost.

Operational Procedures of Incubation

The working relationship between incubators and incubated startups or individuals (referred to as talent) is governed by agreed-upon terms, which outline how the incubation will be conducted. These terms typically include:

- Minimum working hours and project evaluation timelines.
- Behavioral expectations.
- Duration of incubation, often ranging from 18 to 24 months (with possibility of extension).
- Limits on use of resources and infrastructure.
- Rights, responsibilities, and other incubation-specific conditions.

Most TBIs adopt a "resource-first, hands-off" approach — they provide access to essential tools and support but allow incubated teams to independently manage their development processes. Once the talent is briefed on available resources, they are free to utilize them as needed, promoting autonomy and creativity without rigid work routines.

Education and Learning Opportunities

In addition to infrastructure and services, TBIs emphasize continuous learning. They organize specialized training sessions led by industry experts and facilitate networking opportunities with venture capitalists. These educational interventions serve dual purposes:

- Ensuring that incubated teams enhance their skills and develop stronger products or services.
- Enhancing the incubator's reputation within the broader business and investment ecosystem.

Performance Review and Appraisal

To track progress and maintain standards, TBIs conduct at least biannual evaluations of their incubated talent. These reviews are performed by a committee that includes incubator staff, selection panel members, and industry professionals. The review not only assesses the performance and outcomes of the incubated startups but also evaluates the effectiveness of the incubator's own support services. The feedback from this process is used to improve the quality and impact of the incubation experience.

Formal entry: Once validated, the startup officially joins the incubator program, gaining access to its resources and support services.

Shared workspace: Incubators offer shared office space, meeting rooms, and other facilities, reducing overhead costs for startups.

Mentorship and coaching: Experienced mentors and advisors provide continuous guidance on various aspects like financial management, marketing, and legal matters.

Networking opportunities: Incubators facilitate connections with potential investors, partners, and customers, expanding the startup's network and access to resources.

Training and workshops: Startups can participate in training programs and workshops to gain essential skills in areas like business development, marketing, and fundraising.

3. Graduation and Post-Incubation Process

Graduating from an Incubator

The process of finishing incubation is traditionally known as 'graduating' and most incubators have differing criteria for gauging when a talent can graduate from an incubator and what that entails from both the incubator as well as the talent. Some of these criteria include self- sustenance in terms of revenue stream, selling the venture to a larger company, expiry of the period specified in the terms of agreement, etc. These conditions are also known to be multifaceted to ensure that the talent is adequately charged (if at all) for the services it avails at the incubator.

For example, if a company is acquired straight out of the incubator then it is quite common for the talent to encash certain amount of equity for the incubator. However, if the talent is going mainstream by themselves, terms of deferred encashment of equity or even free services are common to allow the company to survive their initial years as an independent entity.

Exit strategy: The incubator and startup work together to develop a suitable exit strategy, which could involve graduation to independent operation, acquisition by another company, or a pivot to a different business model.

Continued support: Incubators may offer ongoing support and resources even after graduation, such as alumni networks and access to specialized services.

Performance monitoring: Incubators often track the progress of graduated startups and measure their impact on the broader economy and ecosystem.

Profit-oriented incubators

Profit-oriented incubators are business development programs that aim to generate financial returns for their investors or owners. Unlike nonprofit incubators, which often focus on social or economic impact, profit-oriented incubators prioritize commercial success and financial gain.

Key Characteristics of Profit-Oriented Incubators

- **Financial Investment:** They typically invest capital in startups in exchange for equity or other financial interests.
- **Revenue Generation:** Their primary goal is to generate profits through the growth and success of the incubated companies.
- **Hands-On Support:** They often provide active mentorship, guidance, and resources to help startups achieve commercial viability.
- **Exit Strategy:** They have a clear exit strategy, such as selling their equity stake or merging with another company, to realize financial returns.

Examples of Profit Oriented Incubators

- National Agricultural Science Technology Incubator (NASTI): A government-funded incubator that supports agricultural technology startups
- **AgFunder:** A global agricultural technology investment platform that connects investors with startups in the agriculture and food sector.
- **Seedfund:** An early-stage venture capital fund and incubator that invests in agricultural technology startups.

Advantages

- **Financial Incentives:** Profit-oriented incubators can provide significant financial resources to startups.
- **Experienced Mentorship:** They often have experienced entrepreneurs and investors who can offer valuable guidance.
- **Focus on Commercial Success:** Their focus on profitability can help startups develop sustainable business models.

Disadvantages

- **Profit-Driven Approach:** The emphasis on financial returns may sometimes overshadow other important factors, such as social impact or long-term sustainability.

- **Limited Support:** Compared to nonprofit incubators, profit-oriented incubators may have fewer resources and support services.

Syngenta Case Study on Sunflower Production

Syngenta uses cutting-edge breeding methods to develop high performing, multi-disease tolerant hybrids which are better for rainfed conditions. Five years ago Tirupathama and her husband M Swamulu, from Kurnool, Andhra Pradesh, were dealing with huge debts and could hardly make ends meet. Because of this reason, more than a decade they lived below the poverty line due to inadequate returns. In 2009 they started using Syngenta's seed SB-207 along with its protocols for sunflower. They cultivated the hybrid on three acres of land investing Rs. 12000 in one season. Their net income was Rs 69,000 which was one of the highest returns in their village Chotukuru.

Non-profit Oriented Incubators

Non-profit oriented incubators are technology incubation programs that prioritize social and economic impact over financial returns. They typically focus on supporting startups that address societal challenges or promote sustainable development.

Key Characteristics of Non-Profit Oriented Incubators

- **Social Mission:** They have a clear social or environmental mission, and their programs are designed to achieve specific social or environmental goals.
- **Impact Measurement:** They use impact measurement tools to track the social and environmental benefits of their programs.
- **Community Engagement:** They often have strong ties to local communities and work to promote social and economic development in their region.
- **Philanthropic Funding:** They rely on philanthropic funding, government grants, or donations to support their operations.
- **Collaborative Approach:** They often collaborate with other non-profit organizations, government agencies, and businesses to achieve their goals. Examples of non-profit oriented incubators
- **Social Enterprise Incubators:** These incubators focus on supporting startups that address social or environmental problems while generating revenue.

- **Clean Technology Incubators:** These incubators support startups developing clean energy technologies and sustainable products.
- **Health Technology Incubators:** These incubators support startups developing medical devices, pharmaceuticals, or healthcare services.
- **Community-Based Incubators:** These incubators are located in rural or underserved communities and aim to promote economic development and job creation.

Advantages of Non-Profit Oriented Incubators

- **Social Impact:** They can have a significant positive impact on society and the environment.
- **Community Development:** They can promote community development and economic growth in underserved areas.
- **Philanthropic Support:** They can attract philanthropic funding and support from individuals and organizations.
- **Collaborative Approach:** They can leverage partnerships with other organizations to achieve their goals.

Green Energy Innovations: SELCO India

SELCO India, a social enterprise incubated in Bangalore, focuses on renewable energy solutions for underserved communities. Founded by Harish Hande, SELCO has pioneered solar energy access in rural areas. Their case study highlights how customized solar systems can improve livelihoods. For Example, Rajesh, a farmer in Karnataka, installed a solar-powered irrigation pump. This not only increased his crop yield but also reduced his dependence on diesel generators. SELCO's impact extends beyond energy—it empowers communities to thrive sustainably.

Technology Business Incubators in India: An Overview

Technology Business Incubators (TBIs) have been part of India's innovation ecosystem since the 1980s, initially backed by the Government and later supported by the private sector from the late 1990s onwards. They have been instrumental in promoting risk-taking, particularly in the information technology sector, and fostering innovation through public research. This early success led to a rapid expansion, resulting in more than 300 officially registered incubators operating across the country by early 2014.

The Government of India, primarily through the **Department of Science & Technology (DST)**, provides structural support to encourage the growth of technology-driven and knowledge-based enterprises. This initiative is led by the **National Science & Technology Entrepreneurship Development Board (NSTEDB)**, which includes representatives from various socio-economic and scientific departments. NSTEDB aims to transform individuals from being "job seekers" to "job creators" using Science & Technology (S&T) interventions and programs.

Core Objectives of NSTEDB

NSTEDB's goals include

- Promoting high-level entrepreneurship among individuals with S&T backgrounds and enabling self-employment by utilizing existing infrastructure and scientific methods.
- Offering support services and information to encourage entrepreneurial ventures.
- Connecting academic, R&D institutions, and support agencies to promote entrepreneurship, particularly in underdeveloped areas.
- Advising on policies related to entrepreneurship and enterprise development.

To achieve these goals, NSTEDB has implemented two major initiatives:

1. **Science & Technology Entrepreneurs Parks (STEP)** – initiated in the 1980s.
2. **Technology Business Incubators (TBI)** – launched in the early 2000s.

Both initiatives have contributed significantly to nurturing start-ups and encouraging innovation, while also supporting economic development at both local and national levels.

Purpose and Role of TBIs in Emerging Economies

In developing countries like India, TBIs are designed with a dual purpose: driving economic growth and enhancing research and innovation. Their role is to support a knowledge-based economy that can compete globally, while still nurturing local talent and technological development. Over the past decade, TBIs have been set up in prominent institutions across India—ranging from business schools to scientific research universities.

According to NSTEDB, the main objectives of TBIs include

- Supporting the creation of new technology-based enterprises.
- Generating skilled employment and value-added services.
- Enabling effective technology transfer.
- Promoting entrepreneurial culture.
- Accelerating the commercialization of research outcomes.
- Providing specialized services to existing small and medium enterprises (SMEs).

Impact and Future of TBIs

The TBI model has seen notable success in India, leading to the emergence of private incubators that operate similarly but with a greater focus on private investment returns (e.g., leveraging platforms like Microsoft Azure). These incubators continue to promote innovation, job creation, and wealth generation, contributing meaningfully to the broader economy.

Recognizing this success, the Indian Government allocated nearly **$2 billion USD** in the **2014–2015 budget** to support incubators and jointly fund promising start-ups alongside private investors. This major investment underscores the Government's commitment to fostering innovation through initiatives like **'Make in India'**, and highlights the key role TBIs have played in shaping India's entrepreneurial ecosystem.

Technology Business Incubators in India

1. Centre for Innovation, Incubation and Entrepreneurship (CIIE), IIM Ahmedabad
2. Technopark-TBI-KOCHI
3. TBI, Birla Institute of Technology, Pilani
4. Rural Technology and Business Incubator, IIT-Madras
5. Society for Innovation and Entrepreneurship, IIT Bombay
6. Microsoft Ventures
7. Startup Village

List of Science & Technology Entrepreneurship Parks (STEPs)/ Technology Business

Incubators (TBIs) recognized by the Government of India

Sl.No.	Name and Address of Contact Person - TBI
1	Amity Business Incubator E-3 Block ,Ist Floor, Sector 125, Amity University Campus, Noida
2	Society for Development of Composites Composites Technology Park 205, Bande Mutt, Kengeri Satellite Township, Bangalore - 560060
3	Technopark – Technology Business Incubator Trivandrum 695 581
4	Society for Innovation and Entrepreneurship Indian Institute of Technology-Bombay Powai, Mumbai 400 076
5	Vellore Institute of Technology (VITTBI) Vellore - 632014
6	Technology Business Incubator – University of Madras Taramani campus, Chepauk, Chennai 600113.
7	Rural Technology & Business Incubator Indian Institute of Technology Madras Chennai 600036
8	Bannari Amman Institute of Technology – Technology Business Incubator Sathyamangalam - 638 401.
9	Periyar Technology Business Incubator Periyar Maniammai College of Technology for Women.
10	JSSATE – Science and Technology Entrepreneurs' Park J.S.S. Academy of Technical Education, C-20/1, Sector-62, Noida-201301, (U.P).
11	Krishna Path Incubation Society Krishna Institute of Engineering & Technology 13 KM Stone, Ghaziabad - Meerut Road, Ghaziabad 201206
12	Entrepreneurship Development Center, NCL Innovation Park, Pune.
13	SJCE – STEP S.J. College of Engineering, Mysore - 570 006
14	Centre for Innovation Incubation and Entrepreneurship (CIIE) Indian Institute of Management, Vastrapur Ahmedabad 380015
15	NITK - Science & Technology Entrepreneurs Park National Institute of Technology – Karnataka Surathkal 575025
16	NITK - Science & Technology Entrepreneurs Park National Institute of Technology – Karnataka Surathkal 575025
17	Science and Technology Park University of Pune, Pune – 411007
18	Science & Technology Entrepreneurs Park - Thapar University Patiala -147001 Punjab
19	TREC-STEP TREC-STEP, NIT Campus Tiruchirappalli 620015
20	PSG-STEP PSG College of Technology, Peelamedu Coimbatore 641004 Tamil nadu
21	STEP - Indian Institute of Technology, Kharagpur - 721 302.
22	STEP - Guru Nanak College of Engineering, Ludhiana

Schemes for Promoting Incubators in India

1. **Ministry of Micro, Small and Medium Enterprises (MoMSME) Schemes**
   a. Scheme for Promoting Innovation, Rural Industries and Entrepreneurship (ASPIRE)
   b. Support for Entrepreneurial and Managerial Development of MSMEs through Incubators
   c. Scheme for Setting up for NSIC Training-cum-Incubation Centre (NSIC-TIC) for Small Enterprise Establishment under PPP Model
   d. Building Awareness on Intellectual Property Rights (IPR) for Micro, Small and Medium Enterprises
   e. Udyam Sakhi
2. **Ministry of Science and Technology: Department of Science and Technology (DST) Schemes**
   a. National Initiative for Developing and Harnessing Innovations Technology Business Incubator (NIDHI-TBI)
   b. National Initiative for Developing and Harnessing Innovations Seed Support System (NIDHI-SSS)
   c. National Initiative for Developing and Harnessing Innovations Entrepreneur-In-Residence (NIDHI-EIR)
   d. National Initiative for Developing and Harnessing Innovations Promotion and Acceleration of Young and Aspiring Technology Entrepreneurs (NIDHI-PRAYAS)
   e. National Initiative for Developing and Harnessing Innovations Grand Challenges and Competitions for Scouting Innovations (NIDHI-GCC)
   f. National Initiative for Developing and Harnessing Innovations Centres of Excellence (NIDHI-CoE)
   g. National Initiative for Developing and Harnessing Innovations-Accelerator (NIDHI Accelerator)
   e. NewGen Innovation and Entrepreneurship Development Centre (NewGen IEDC)

3. **Ministry of Electronics and Information Technology (MeitY) Schemes**
    a. Technology Incubation and Development of Entrepreneurs (TIDE) Scheme
    b. Support for International Patent Protection in Electronics and IT (SIP-EIT) II
    c. Scheme to Support IPR Awareness Workshop/Seminars in E & IT Sectors
4. **National Institution for Transforming India (NITI Aayog)**
    a. Atal Innovation Mission (AIM)
    b. Women Entrepreneurship Platform (WEP)
5. **Stand-up India**
6. **Innovations for Defence Excellence (iDEX) Initiative**
7. **Bioincubators Nurturing Entrepreneurs for Scaling Technologies (BioNEST)**

Government-backed Programs

Startup India Seed Fund: Provides equity-free grants of up to INR 20 lakhs to eligible startups to validate their ideas.

Startup India Stand-Up India Scheme: Supports incubators catering to women and SC/ST entrepreneurs by providing grants and incubation space.

Fund for Startup Incubation (FSI): Offers financial assistance to eligible incubators across sectors, helping them upgrade infrastructure and expand their capabilities.

Startup Research and Innovation Grant (SRIG): Promotes innovation through collaborative research projects between startups and academic institutions.

Innovation Cell Scheme (ICS): Encourages innovation within educational institutions by establishing and supporting student-led incubators.

Private and Partnership Initiatives

Angel Investor Networks: Connect startups with potential investors, facilitating access to crucial funding.

Co-working Spaces and Hubs: Offer flexible workspaces and networking opportunities for entrepreneurs.

Accelerators and Mentorship Programs: Provide intensive support and guidance to high- potential startups over a fixed period.

Corporate Social Responsibility (CSR) Initiatives: Many companies actively support incubators and startups through CSR programs.

5

Technology Promotion and Essential Skills for Technology Commercialisation

Technology Promotion

The activity of promoting the transfer of technology by creating opportunities, advertising and organizing technology events.

Promotion is a marketing tool, used as a strategy to communicate between the sellers and buyers. Through this, the seller tries to influence and convince the buyers to buy their products or services. It assists in spreading the word about the product or services or company to the people. The company uses this process to improve its public image. This technique of marketing creates an interest in the mindset of the customers and can also retain them as a loyal customer.

Promotion is a fundamental component of the marketing mix, which has 4 Ps:

1. Product (Warranty, features, variety, quality, design, packaging, brand name)
2. Price (List price, discounts, payment terms, credit)
3. Place (channels, coverage, Types of promotion locations, transportation, inventory)
4. Promotion (advertising, self-marketing/personal, sales promotion, public relations, direct marketing, trade shows, events).

Types of Promotion

1. Advertising

The activity of making products or services known about and persuading people to buy them. It helps to outspread a word or awareness, promote any newly launched service, goods of an organization. The company uses advertising as a promotional tool as it reaches a mass of people in a few seconds. An advertisement is communicated through many traditional media such as radio, television, outdoor advertising, newspaper. Other contemporary media that supports advertisement are social media, blogs, text messages and websites.

2. Sales Promotion

Short-term campaigns to spark interest and create demand for a product or service. Short term incentives which are offered to the ultimate customers to encourage them to make immediate purchase and increase sales. However, it is for a limited time, used to expand customers demand, refresh market demand and enhance product availability. Ex: Contests, coupons, freebies, buy 1 get 1 free, free shipping, flash sales. Used when

a) The product is newly introduced. b) There is excess inventory.

c) Penetration or entry into a competitor's stronghold.

d) The products are likely to be perished if not sold or used e) To draw attention, buy 1 get 1 free

3. Personal/Self-promotion

It is a process where the enterprises send their agents directly to the customers to pitch for their product or service (Or) personal promotion, also known as self-promotion, is the act of presenting yourself to others to showcase your skills, experience, and accomplishments. It can be done in many ways, including:

Speaking: In public speeches or face-to-face conversations Social media: On blogs or social media platforms Networking: At events or through leadership campaigns Marketing: For a product or service

Posture, speech, or dress: Through your mannerisms, posture, speech, or dress. Self-promotion can help an individual in:

Stand out: Differentiate yourself from other candidates during an interview

Improve your career: Increase ur chances of a promotion or raise, or new opportunities

Build credibility: Establish yourself as an expert in your industry

Mentor others: Support your colleagues and help them improve their skills

Build your personal brand: and broadcast your name

4. Online Promotion

Includes almost all the elements of the promotion mix. Starting from the online promotion with pay per click advertising. Direct marketing by sending newsletters or emails. It is marketing products and services using the internet. It can include a mix of strategies, such as:

Social media marketing: A cost-effective way to connect with target audience, increase brand recognition, and improve customer experience.

Email marketing: A way to build and utilize your email list.

Google Ads: A way to advertise ur products and services.

Sponsorships: A way to demonstrate your values and get your name in front of potential customers.

Limited-time offers: A way 2convert more prospects by creating a targeted pitch.

Personalized promotions: A way to personalize your promotions and maximize your net profit.

5. Public Relation

Is the practice of managing and sharing information with the public to influence their perception of a brand, company, or organization. The goal of PR is to build mutually beneficial relationships between the organization and the public. It is exercised to broadcast the information or message between a company (NGO, Government agency, business), an individual or a public. A powerful PR campaign can be valuable to the company. It helps in

Building relationships: with key stakeholders across various platforms.

Maintaining a positive image: for a company, especially for publicly traded companies. Handling media requests: and shareholder inquiries.

Using media relations: PR uses media relations to earn press coverage. Using content: PR uses content to measure performance.

6. Direct Promotion/ Direct Marketing

It is that kind of advertising where the company directly communicates with its customers. This communication is usually done through various new approaches like email marketing, text messaging, websites, fliers, online adverts, promotional letters, catalogue distributors, House-to-house selling, Shop-at-home services, Direct mail, Telemarketing, Social media ads etc. It's a sales communication technique that doesn't use any intermediary media. It helps companies develop strong relationships with buyers and sell products.

Key Points of Promotion

- **Understanding your audience:** Consider who you're promoting the technology to and what their concerns are, convince customers to buy goods and services. The result or outcome of the promotion is immediate.
- **Highlighting benefits:** Emphasize the benefits of the technology.
- **Engaging and educating:** Engage with your audience and educate them about the technology.
- **Leveraging influencers:** Consider working with influencers to promote the technology. It can be used for all sorts of businesses irrespective of the size, brand of a company
- **Marketing:** Marketing is important for the development of new technologies, both before and after the product is on the market. It is an economic marketing tool
- **Social media:** Social media is a key component of tech PR strategies, helping to build brand awareness and increase engagement with customers.
- **Analytics:** Analytics can help you evaluate the effectiveness of your marketing strategies and plan for the future of your brand.

Technology promotion encompasses a broad range of activities and initiatives aimed at accelerating the adoption and utilization of new technologies to achieve economic, social, and environmental benefits.

Overall, technology promotion plays a crucial role in bridging the gap between innovation and its real-world application. By employing a diverse range of activities and engaging various stakeholders, technology promotion can drive economic growth, create jobs, solve societal challenges, and improve the quality of life for people around the world.

Here's a closer look at different types of technology promotion activities:

1. Business Meetings

A business meeting is a gathering of two or more people to discuss business related matters and take decisions. It is a verbal form of communication among people. Business meetings are more useful than other communication tools such as e-mail, chat or share point. Meetings are a way of getting a group together to discuss a common issue. one side needs something from the other. In a business meeting, both sides benefit mutually.

Matchmaking events: Connecting technology providers with potential users from industry, business, and government sectors.

Investment forums: Providing platforms for startups and technology companies to pitch their ideas to investors and venture capitalists.

Delegation visits: Facilitating exchanges between researchers, entrepreneurs, and industry experts from different countries or regions.

Need of Business Meetings in technology promotion

- To make the decisions regarding the technology
- Sharing of information related to technology
- To announce the changes in the technology
- Identifying the partners or potential buyers

Purpose: To discuss operations, address changes, celebrate successes and make decisions

Participants: Employees, management, stakeholders, external partners

Structure: Usually follow a clear structure dictated by their objectives

Types: Team building, innovation, information sharing, status update, decision-making, problem-solving.

2. Scientist-Industry/Entrepreneur Meets

Scientist –industry meet is defined as a group of scientists and entrepreneurs meet to discuss regarding business related matters.

It involves multiple goals like Updating about research progress, providing opportunities for collaboration, and Facilitating discussion to get everyone involved.

Role of scientist- industry/ entrepreneur meeting in technology promotion

- Introducing of the technology by scientist
- Features & Importance of the product
- Value of the product in future
- Details of patent if any

- Identifying the potential partners for commercialization
- To create the interest among entrepreneurs

Knowledge transfer workshops: Enabling scientists and researchers to showcase their innovations and share expertise with potential industry partners.

Joint research and development projects: Fostering collaboration between academia and industry to commercialize research findings and translate them into practical applications.

Innovation challenges: Inspiring researchers and entrepreneurs to develop solutions to specific technological challenges faced by industry.

3. Technology Conclave

It is an event that brings together people to discuss and learn about technology. The event provides a platform for various stakeholders to learn about new technologies and applications. The event will bring together industry leaders, policymakers, and academia.

Large-scale conferences: Bringing together key stakeholders from industry, academia, government, and non-profit organizations to discuss trends, challenges, and opportunities in specific technology sectors.

Exhibitions and demonstrations: Showcasing cutting-edge technologies and providing opportunities for networking and knowledge exchange.

Awards and recognition programs: Recognizing and celebrating innovation and technological achievements.

4. Business Plan Competition

Encouraging entrepreneurship: Providing platforms for entrepreneurs to pitch their business ideas to a panel of judges and secure funding or mentorship.

Identifying promising ventures: Helping investors and industry partners discover high- potential startups with innovative technology solutions.

Nurturing business skills: Providing feedback and guidance to entrepreneurs on refining their business plans and improving their presentation skills.

5. Farmers Fairs

Farmer's fair means an exhibition that is intended to promote agriculture by including a balanced variety of exhibits of livestock and agricultural products, as well as related manufactured products.

Farmer's fair is one of the most important extension education activities of the university for promotion and transfer of technology to the potential users/ farmers. Highlights the importance of modern agricultural techniques and the role of progressive farmers in leading agricultural development. They include display of product, demonstrations.

Disseminating agricultural technology: Showcasing new agricultural technologies, tools, and best practices to farmers and rural communities.

Promoting knowledge sharing: Facilitating exchange of knowledge and experiences among farmers through demonstrations, workshops, and farmer-to-farmer interactions.

Improving agricultural productivity: Encouraging farmers to adopt new technologies to improve yields, income, and resource efficiency.

6. Technology Shows

A technology show is an event that showcases technology along with promotion and can include a variety of activities, such as

Technology showcase: A place where customers can see a company's new products. Technology showcases can also include networking sessions, lectures, and corporate fairs for promoting their product

Technology fair: An interactive event that promotes the transfer of technology. Technology fairs can include exhibition booths, interactive activities, and side events.

Tech expo: A business-to-business trade show that allows companies to network, promote their businesses, and display their work.

Public awareness and engagement: Showcasing advancements in various technology sectors to the general public, sparking interest and encouraging adoption.

Interactive exhibits and demonstrations: Providing hands-on experience with new technologies to foster understanding and appreciation.

Promoting responsible technology use: Raising awareness about potential benefits and risks of new technologies and encouraging responsible development and deployment.

Dealing with Entrepreneurs, Agripreneurs and Other Stakeholders

Business Communication

Communication is neither the transmission of a message nor the message itself. It is the mutual exchange of understanding, originating with the receiver. Communication needs to be effective in business. Communication is the essence of management. The basic functions of management (Planning, Organizing, Staffing, Directing and Controlling) cannot be performed well without effective communication.

Business communication involves constant flow of information. Feedback is integral part of business communication. Organizations these days are very large and involve large number of people. There are various levels of hierarchy in an organization. Greater the number of levels, the more difficult is the job of managing the organization.

Communication here plays a very important role in process of directing and controlling the people in the organization. Immediate feedback can be obtained and misunderstandings if any can be avoided. There should be effective communication between superiors and subordinated in an organization, between organization and society at large (for example between management and trade unions). It is essential for success and growth of an organization. Communication gaps should not occur in any organization.

Effective communication: Clearly and concisely conveying your message through the appropriate channels (email, phone, meetings, presentations) and tailoring your communication style to your audience.

Active listening: Paying close attention to others, understanding their needs and perspectives, and ensuring clear two-way communication.

Professional written and verbal communication: Using proper grammar, vocabulary, and tone, avoiding jargon and slang, and maintaining clarity and professionalism in all your interactions.

Persuasive communication: Presenting your ideas effectively, building logical arguments, and adapting your approach to different audiences.

Business Communication is goal oriented. The rules, regulations and policies of a company have to be communicated to people within and outside the organization.

Business Communication is regulated by certain rules and norms. In early times, business communication was limited to paper-work, telephone calls etc.

But now with advent of technology, we have cell phones, video conferencing, emails, satellite communication to support business communication. Effective business communication helps in building goodwill of an organization.

Business Communication can be of two types

1. **Oral Communication** - An oral communication can be formal or informal. Generally business communication is a formal means of communication, like : meetings, interviews, group discussion, speeches etc. An example of Informal business communication would be - Grapevine.
2. **Written Communication** - Written means of business communication includes - agenda, reports, manuals etc.

Business Etiquette

Business etiquette is a set of rules that govern the way people interact with one another in business, with customers, suppliers, with inside or outside bodies. It is all about conveying the right image and behaving in an appropriate way.

Rules for good business etiquette

1. Always use Names in a Meeting

It is easy to forget people's names when in a business meeting, and for this reason, it is a good idea to write all the names down (and check their spelling) on a piece of paper in front of you.

It is common to say 'treat others as you would like to be treated yourself'. However, different people have different expectations. You might like to be called by your first name when being greeted by a service provider, however, someone else might prefer to be referred to more formally, with Mr, Ms, or Mrs.

You should not just assume that people like things the same way you do. If in doubt about the level of formality to use, it is probably better to go for more formality, rather than less, to avoid offending anyone.

2. The three Rs

It is important to be considerate about the psychological needs of different people. A very useful rule of thumb to go by is that of the three Rs.

- **Recognition:** using names, greetings, and making a point of acknowledging people.

- Respect: treating people with respect, value and courtesy, and apologising to them where the situation calls for it.
- Response: people do not want to be kept waiting, they need to be responded to.

3. Wardrobe and Hygiene

Dirty clothes, fingernail biting, poor hygiene, unclean hair and body odour especially, can be a real turn off. It can be difficult having to tell someone they have body odour, but it is necessary, especially if the person has to deal with others and outside clients.

4. Cracking Inappropriate Jokes

Etiquette is all about behaviour and sensitivity. For example, cracking jokes at times where it is inappropriate, or inappropriate jokes in general, shows a total lack of sensitivity.

I have been surprised to still hear some men at work make jokes about women and the way they look. Also, jokes about race and disabilities are most certainly inappropriate at any time.

5. Showing Gratitude

Thanking a person where the thank you is warranted is simple politeness.

Actually making a point of showing some sort of gratitude where someone has gone out of their way for you, or performed a task that is not part of their job description, is very important and actually makes for better interpersonal communication in the future.

6. Telephone Etiquette

Being treated rudely on the telephone or left on hold is not professional. Making promises and then not keeping them or following through is also unprofessional and projects a poor image to the people on the receiving end.

How to improve business etiquette

Often people do not even realise they are not showing a high level of etiquette. The best way to develop good business etiquette would be to get together as a team and consider:

- How do we work with one another?
- What is the image we convey to people?

- Are we polite enough?
- How can we show more respect, be more responsive and recognise people?
- What are our standards?

Often you can generate improvements in just that one session. Rather than calling it 'business etiquette', you could call it 'customer service improvement' or 'responsiveness'.

Business etiquette can sound like it focuses on small things that are unimportant, but all together they make a big difference to the workplace and the kind of responses received in all areas.

Professional attire: Dressing appropriately for your specific business context, presenting a polished and respectful image.

Punctuality and time management: Arriving on time for meetings, deadlines, and appointments, demonstrating respect for others' time.

Proper greetings and introductions: Using appropriate salutations, introductions, and titles when addressing colleagues, clients, and superiors.

Non-verbal communication: Maintaining good posture, eye contact, and a positive demeanor, conveying professionalism and confidence.

Cultural awareness: Being mindful of cultural differences and adjusting your communication and behaviour accordingly when interacting with international colleagues or clients.

Business Networking

Business networking is a way of leveraging your business and personal connections to help you bring in new customers, vendors or to get great advice for running your business.

Business networking is a term that refers to meeting other business owners, potential suppliers, or other professionals who have business experiences—to help you grow your business. Networking gives you a pool of experts that range from competitors to clients, and allows you to offer something to them; hopefully in exchange for their services, advice, knowledge, or contacts.

Developing relationships as a business owner and offering assistance to others does more than give you potential clients or generate referrals. Networking assists you in identifying opportunities for partnerships, joint ventures, or new areas of expansion for your business.

How Does Business Networking Work?

Networking events or local business luncheons present you with opportunities to find others who are in similar circumstances as you work to grow your business. These events are generally put together for the purpose of introducing new concepts and methods being used while providing a platform for local business people to meet and exchange ideas.

When you meet someone, be sure to exchange business cards, and follow up later discussing points or topics brought up in conversations you may have had with them. After a few conversations, you may be able to bring up the issues you are facing. If they open up discussions first, you might be able to begin exchanging information, seeking knowledge, or exchanging business contacts.

Most business people are optimistic and positive. Regular association with such people can be a great morale boost, particularly in the difficult early phases of a new business. You'll find that many, if not all, business owners have experienced similar trials of ownership.

Much of local business is still done on a handshake basis, and the best way to network with other local business owners and entrepreneurs is through face-to-face meetings and local business groups.

What Are the Benefits of Business Networking?

Business networking might allow you to create awareness of or keep abreast of the latest trends or technology in your industry. A network can also provide you with professional mentors or contacts who might be able to assist you with problems you might need help with.

For example, if your business needs the services of a bookkeeper, accountant, or lawyer you may find the ideal candidate through your network. If your business needs equity financing you may be able to find an angel investor or venture capitalist through networking channels.

Networking is ideal for expanding your knowledge by taking advantage of the viewpoints and prior experience of others. For instance, if you are thinking of exporting your products or services, you may be able to get some valuable advice from someone else who has done similar business internationally.

Types of Business Networking

As you attend events, look for indicators that someone might be in a position to benefit your business, where you have something to offer as well. This could simply be conversations about your industry's market conditions as well as the trends each of you may have noticed within your industry. You can work together to develop an understanding of the market you both operate in.

Business Seminars

Look for and attend some business seminars—cultivate new working relationships with your new peers and business associates, then communicate on a regular basis to help you all stay current. The most important skill for effective business networking is listening; focusing on how you can help the person you are listening to rather than on how they can help you is the first step to establishing a mutually beneficial relationship.

Networking Groups

The best business networking groups operate as exchanges of business information, ideas, and support. There are many groups online that offer networking services and communities— LinkedIn is an example of a large networking group or site that can bring professionals together.

Professional Associations

Organizations exist that are comprised of like-minded individuals in similar industries and fields of work. These organizations may have entry fees or other selection requirements, but they can prove vital for small business owners looking to expand their network. The American Management Association and the American Marketing Association are two examples of industry-specific associations.

Building relationships: Actively seeking opportunities to connect with colleagues, clients, and potential partners in your field.

Attending industry events and conferences: Participating in relevant conferences, workshops, and networking events to expand your professional network.

Making a good first impression: Presenting yourself confidently, introducing yourself effectively, and engaging in meaningful conversations with new contacts.

Following up and maintaining relationships: Staying in touch with connections after initial contact, sending thank-you notes, and providing value to your network.

Utilizing online networking platforms: Leveraging platforms like LinkedIn to connect with professionals in your industry and build your online presence.

Interconnectedness of these elements

Effective business communication is essential for building trust and rapport in networking situations.

Understanding business etiquette demonstrates professionalism and makes networking interactions more comfortable and productive.

Networking expands your professional circle, providing access to information, opportunities, and potential collaborations, where strong communication skills are vital for success.

6

Emerging Approaches in Technology Commercialisation and Incubation

Technology Scouting and Innovations in Technology Incubation

Technology scouting and innovations in technology incubation have a symbiotic relationship, fuelling each other for the benefit of budding startups and the broader technology ecosystem. Let's dive deeper into their individual roles and how they interact:

Technology Scouting

Technology Scouting is the process of discovering, analyzing and evaluating new or existing technologies that will help them with their innovation process. In simple terms, it is a process for companies to find the necessary technology outside the company.

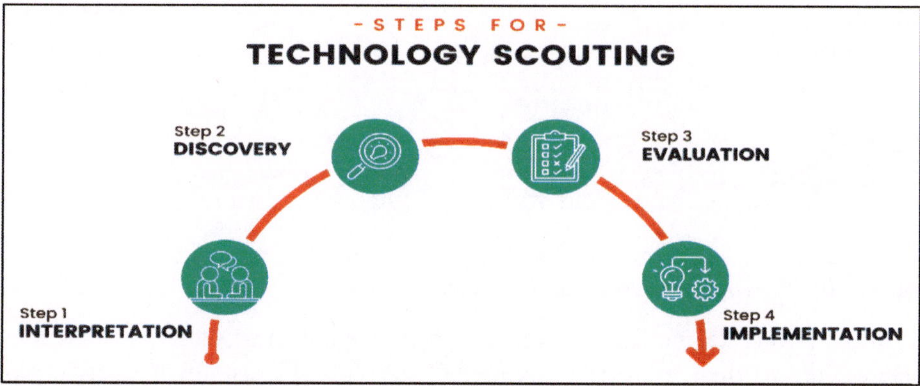

Hunting for promising technologies: Scouts actively search for emerging technologies, patents, research advancements, and market trends across various sectors.

Evaluating potential and fit: They assess the technical feasibility, market potential, and compatibility with the needs of incubators and their supported startups.

Connecting innovators and ventures: Scouts bridge the gap between technology creators (researchers, inventors, startups) and potential users (incubators, established companies).

Fuelling innovation pipelines: Technology scouting helps incubators access a wider range of technologies and ideas, enriching their portfolio and diversifying the innovations supported.

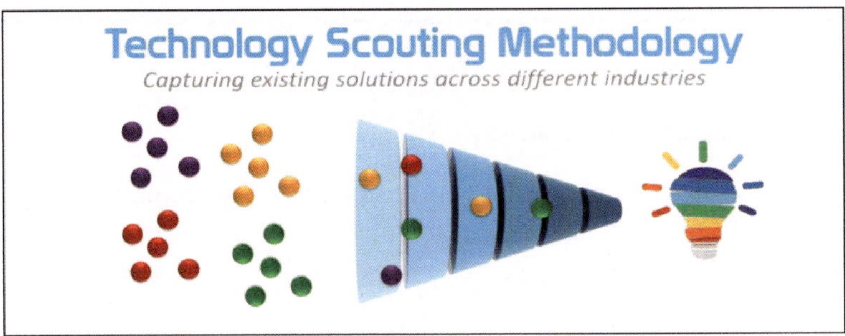

A case study on the Deutsche Telekom Laboratories

Process of Technology Scouting at DTAG

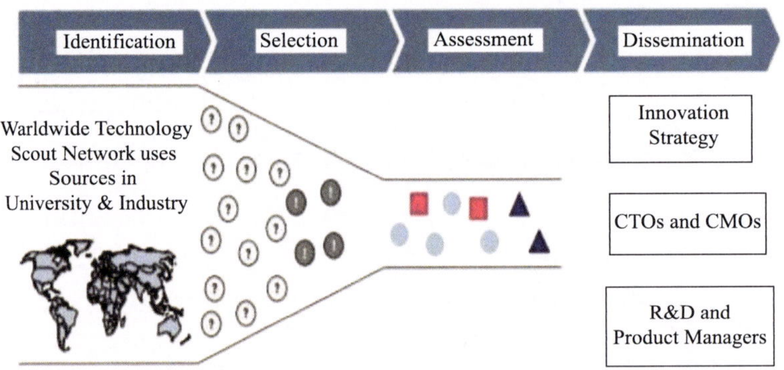

Source: The Technology Radar – Edition II/2006

In order to create monitoring and scanning capabilities, which stretch over most of the world, Deutsche Telekom AG has chosen to establish a network of technology scouts.

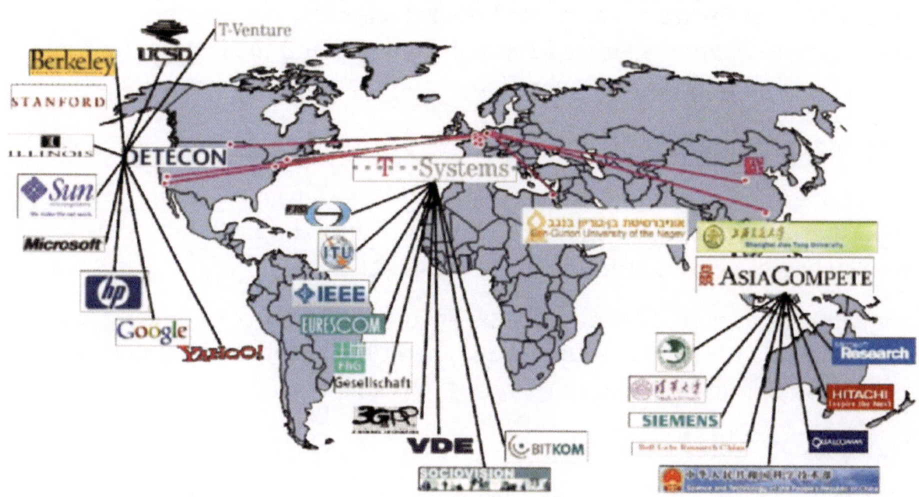

Source: Deutsche Telekom Laboratories (2006), The Technology Radar – Edition II/2006

Innovations in Technology Incubation

Adapting to diverse technologies: Incubators need to develop flexible models and support systems to accommodate the unique needs of startups working with different technologies.

Nurturing cross-disciplinary collaboration: Facilitating interactions between innovators from different fields can spark new ideas and lead to novel applications of existing technologies.

Promoting open innovation: Encouraging collaboration between startups, established companies, and research institutions can accelerate the development and commercialization of new technologies.

Leveraging cutting-edge tools and resources: Implementing advanced scouting methodologies, utilizing data analytics for technology trend identification, and providing access to specialized equipment and facilities can enhance the effectiveness of incubation programs.

How They Interact

Scouting informs incubation: Technology scouting results inform an incubator's selection of startups, focus areas, and strategic partnerships.

Incubation validates scouting: The success of startups using scouted technologies validates the scouting process and identifies promising areas for future exploration.

Feedback loop for improvement: Lessons learned from both successful and unsuccessful ventures, gathered through scouting and incubation, help refine approaches and improve the overall innovation ecosystem.

Examples of synergy

An incubator focused on sustainability might scout green technologies like bioplastics or renewable energy solutions, supporting startups developing these technologies and bringing them to market.

A university-based incubator might use scouting to identify promising research projects with commercial potential and then offer incubation support to those researchers to translate their discoveries into viable ventures.

Examples of successful synergy

DARPA's Advanced Research Projects Agency Network (ARPANET) program: Technology scouting led to the identification of key technologies that formed the foundation of the internet, which was then further developed through incubation efforts.

The Israeli Innovation Authority's incubator programs: These programs have successfully nurtured numerous high-tech startups based on technologies identified through scouting efforts.

Bibliography

Action Plan. Government of India. Retrieved from https://www.startupindia.gov.in/

Aernoudt, R. (2004). Incubators: Tool for entrepreneurship? Small Business Economics, 23(2), 127–135. https://doi.org/10.1023/B:SBEJ.0000027665.54173.23

AgFunder. (n.d.). Investment platform for agritech. Retrieved from https://agfunder.com/

AgrInnovate India Ltd. (AgIn). (n.d.). Technology commercialization and licensing procedures. Government of India. Retrieved from https://agrinnovateindia.co.in/

Allen, D. N., & McCluskey, R. (1990). Structure, policy, services, and performance in the business incubator industry. Entrepreneurship Theory and Practice, 15(2), 61–77.

Alston, J. M., Andersen, M. A., James, J. S., & Pardey, P. G. (2011). Persistence Pays: U.S. Agricultural Productivity Growth and the Benefits from Public R&D Spending. Springer.

Atal Innovation Mission (AIM), NITI Aayog. (n.d.). Atal Innovation Mission Programs. Retrieved from https://aim.gov.in

AUTM (Association of University Technology Managers). (2018). Technology Transfer Practice Manual (4th ed.). AUTM: USA.

AUTM (Association of University Technology Managers). (2022). Technology Transfer Practice Manual. Retrieved from https://autm.net

AUTM. (2020). Technology transfer practice manual. Association of University Technology Managers. Retrieved from https://autm.net

Avnimelech, G., & Teubal, M. (2006). Creating venture capital industries that co-evolve with high tech: Insights from an extended industry life cycle perspective of the Israeli experience. Research Policy, 35(10), 1477–1498. https://doi.org/10.1016/j.respol.2006.09.009

Barras, R. (1990). Interactive innovation in financial and business services: The vanguard of the service revolution. Research Policy, 19(3), 215–237. https://doi.org/10.1016/0048-7333(90)90037-7

Baskaran, A. (2001). Technological innovation and transfer in the Indian pharmaceutical industry. Technology Analysis & Strategic Management, 13(4), 421–445.

Basu, S. (2019). Geographical Indications in India: An Analysis of the Law and Policy. Journal of Intellectual Property Rights, 24(4), 161–170.

Berkes, F. (2012). Sacred ecology (3rd ed.). Routledge.

BIMSTEC Secretariat. (n.d.). Bay of Bengal Initiative for Multi-Sectoral Technical and Economic Cooperation. Retrieved from https://bimstec.org/

Biotechnology Industry Research Assistance Council (BIRAC). (n.d.). BioNEST – Bioincubators Nurturing Entrepreneurs for Scaling Technologies. Retrieved from https://birac.nic.in

Blackwell, R. D., Miniard, P. W., & Engel, J. F. (2006). Consumer behavior (10th ed.). South-Western College Pub.

Blank, S., & Dorf, B. (2012). The Startup Owner's Manual: The Step-By-Step Guide for Building a Great Company. K&S Ranch.

Boettiger, S., & Bennett, A. B. (2006). The Bayh-Dole Act: Implications for developing countries. Innovation Strategy Today, 2(1), 1–18.

Bornstein, D., & Davis, S. (2010). Social entrepreneurship: What everyone needs to know. Oxford University Press.

Bovee, C. L., & Thill, J. V. (2021). Business communication today (15th ed.). Pearson.

Bozeman, B. (2000). Technology transfer and public policy: A review of research and theory. Research Policy, 29(4-5), 627–655. https://doi.org/10.1016/S0048-7333(99)00093-1

Bozeman, B. (2000). Technology transfer and public policy: A review of research and theory. Research Policy, 29(4–5), 627–655. https://doi.org/10.1016/S0048-7333(99)00093-1

Bozeman, B. (2000). Technology transfer and public policy: A review of research and theory. Research Policy, 29(4–5), 627–655. https://doi.org/10.1016/S0048-7333(99)00093-1

Bozeman, B. (2000). Technology transfer and public policy: a review of research and theory. Research Policy, 29(4–5), 627–655. https://doi.org/10.1016/S0048-7333(99)00093-1

Bozeman, B., & Rogers, J. D. (2002). A churn model of scientific knowledge value: Internet researchers as a knowledge value collective. Research Policy, 31(5), 769–794.

Bozeman, B., & Rogers, J. D. (2002). A churn model of scientific knowledge value: Internet researchers as a knowledge value collective. Research Policy, 31(5), 769–794.

Brynjolfsson, E., & McAfee, A. (2014). The Second Machine Age: Work, Progress, and Prosperity in a Time of Brilliant Technologies. W.W. Norton & Company.

Cartagena Protocol on Biosafety. (2000). Text of the Cartagena Protocol. Retrieved from https://bch.cbd.int/protocol/

Cetron, M. (1971). Technological forecasting: A practical approach. Gordon and Breach. Chaffey, D., & Ellis-Chadwick, F. (2019). Digital marketing (7th ed.). Pearson Education.

Chambers, R. (1994). Participatory Rural Appraisal (PRA): Challenges, potentials and paradigm. World Development, 22(10), 1437–1454. https://doi.org/10.1016/0305-750X(94)90030-2

Chambers, R. (1997). Whose Reality Counts? Putting the First Last. Intermediate Technology Publications.

Chambers, R. (1997). Whose reality counts? Putting the first last. Intermediate Technology Publications.

Chaturvedi, A., & Chaturvedi, A. (2011). Business communication: Concepts, cases and applications. Pearson Education India.

Chesbrough, H. W. (2003). Open innovation: The new imperative for creating and profiting from technology. Harvard Business Press.

Chesbrough, H. W. (2003). Open innovation: The new imperative for creating and profiting from technology. Harvard Business School Press.

Christensen, C. M. (1997). The innovator's dilemma: When new technologies cause great firms to fail. Harvard Business Review Press.

Christensen, C. M. (1997). The Innovator's Dilemma: When New Technologies Cause Great Firms to Fail. Harvard Business Review Press.

Christensen, C. M., Raynor, M. E., & McDonald, R. (2015). What is disruptive innovation? Harvard Business Review, 93(12), 44–53. https://hbr.org/2015/12/what-is-disruptive-innovation

Co-operation and Development. Retrieved from https://www.oecd.org

Cohen, B., & Winn, M. I. (2007). Market imperfections, opportunity and sustainable entrepreneurship. Journal of Business Venturing, 22(1), 29–49.

Controller General of Patents, Designs and Trade Marks (CGPDTM). (2020). The Patent (Amendment) Rules, 2020. Ministry of Commerce and Industry, Government of India. Retrieved from https://ipindia.gov.in/

Convention on Biological Diversity (CBD). (1992). Text of the Convention. Retrieved from https://www.cbd.int/convention/text/

Convention on Biological Diversity (CBD). (2000). Article 8(j) - Traditional Knowledge, Innovations and Practices. Retrieved from https://www.cbd.int/traditional/

Council of Scientific and Industrial Research - TKDL Unit. (n.d.). About TKDL. Retrieved April 7, 2025, from https://www.tkdl.res.in

Council of Scientific and Industrial Research. (n.d.). Traditional Knowledge Digital Library (TKDL). Government of India. Retrieved April 7, 2025, from https://www.csir.res.in/tkdl

Czinkota, M. R., & Ronkainen, I. A. (2013). International marketing (10th ed.). Cengage Learning.

Dalrymple, D. G., & Srivastava, J. P. (1994). Public and private sector plant breeding: Thwarting the move toward privatization in the United States and India. World Bank Discussion Paper.

Damodaran, A. (2002). Investment valuation: Tools and techniques for determining the value of any asset (2nd ed.). Wiley.

Damodaran, A. (2012). Investment Valuation: Tools and Techniques for Determining the Value of Any Asset (3rd ed.). John Wiley & Sons.

Das, K. (2006). Protection of Geographical Indications: An Overview of Select Issues with Particular Reference to India. Centre for Trade and Development. Retrieved from https://cuts- citee.org/pdf/Study-Protection_of_Geographical_Indications.pdf

Dasgupta, R., & Sagar, A. (2013). Technology transfer and intellectual property rights: A perspective from India. Technology in Society, 35(1), 35–43. https://doi.org/10.1016/j.techsoc.2013.01.002

Department for Promotion of Industry and Internal Trade (DPIIT). (2016). National Intellectual Property Rights (IPR) Policy. Ministry of Commerce and Industry, Government of India. Retrieved from https://dpiit.gov.in/sites/default/files/National_IPR_Policy_English.pdf

Department of Agriculture and Cooperation. (2002). National Seed Policy. Ministry of Agriculture, Government of India. Retrieved from http://agricoop.nic.in

Department of Agriculture and Cooperation. (2003). Plant Quarantine (Regulation of Import into India) Order, 2003. Ministry of Agriculture, Government of India. Retrieved from http://ppqs.gov.in

Department of Science and Innovation. (2021). Indigenous Knowledge Act No. 6 of 2019. Republic of South Africa. Retrieved from https://www.gov.za/documents/indigenous-knowledge-act-6-2019-english-afrikaans-28-aug-2019-0000

Department of Science and Innovation. (n.d.). Indigenous Knowledge Systems. Republic of South Africa. Retrieved April 7, 2025, from https://www.dst.gov.za/index.php/programmes/indigenous-knowledge-systems

Department of Science and Technology (DST), Government of India. (n.d.). National Initiative for Developing and Harnessing Innovations (NIDHI). Retrieved from https://dst.gov.in

Department of Science and Technology (DST), Government of India. (n.d.). National Science & Technology Entrepreneurship Development Board (NSTEDB). Retrieved from https://dst.gov.in/technology-business-incubation

Deshmukh-Ranadive, J. (2005). Women's Self Help Groups in Andhra Pradesh: Participatory Poverty Alleviation in Action. World Bank.

Dhar, B. (2016). India's New IPR Policy: Myths and Realities. Economic & Political Weekly, 51(22), 13–15. Retrieved from https://www.epw.in/journal/2016/22/commentary/indias-new-ipr-policy.html

Dhar, B., & Rao, N. C. (2002). Protecting Traditional Knowledge: The Indian Experience. Geneva: Quaker United Nations Office.

Dodgson, M., Gann, D., & Salter, A. (2008). The management of technological innovation: Strategy and practice. Oxford University Press.

Drucker, P. F. (1985). Innovation and Entrepreneurship: Practice and Principles. Harper & Row.

Etzkowitz, H., & Leydesdorff, L. (2000). The dynamics of innovation: From National Systems and "Mode 2" to a Triple Helix of university–industry–government relations. Research Policy, 29(2), 109–123. https://doi.org/10.1016/S0048-7333(99)00055-4

Etzkowitz, H., & Leydesdorff, L. (2000). The dynamics of innovation: from National Systems and "Mode 2" to a Triple Helix of university–industry–government relations. Research Policy, 29(2), 109–123. https://doi.org/10.1016/S0048-7333(99)00055-4

Etzkowitz, H., & Zhou, C. (2017). The Triple Helix: University–Industry–Government Innovation and Entrepreneurship. Routledge.

Etzkowitz, H., & Zhou, C. (2017). The Triple Helix: University–Industry–Government Innovation and Entrepreneurship. Routledge.

European Commission. (2015). Guidelines on Intellectual Property Valuation. Retrieved from https://ec.europa.eu

FAO. (2014). Youth and agriculture: Key challenges and concrete solutions. Food and Agriculture Organization of the United Nations.

FAO. (2021). Digital Agriculture Report: Rural E-commerce Development—Experience from China. Food and Agriculture Organization of the United Nations. https://www.fao.org/documents/card/en/c/CB4400EN

Ferrell, O. C., Hirt, G., & Ferrell, L. (2021). Business: A changing world (12th ed.). McGraw-Hill Education.

Fill, C., & Turnbull, S. (2019). Marketing communications: Discovery, creation and conversations (8th ed.). Pearson Education.

Fleming, P. (2020). Business etiquette for dummies (2nd ed.). Wiley.

Food and Agriculture Organization (FAO). (2009). International Treaty on Plant Genetic Resources for Food and Agriculture. Retrieved from https://www.fao.org/plant-treaty/en/

Food and Agriculture Organization. (2019). The State of Food and Agriculture 2019: Moving forward on food loss and waste reduction. FAO.

Food and Agriculture Organization of the United Nations. (2009). International Treaty on Plant Genetic Resources for Food and Agriculture. Retrieved from https://www.fao.org/plant-treaty/en/

Food Safety and Standards Authority of India. (2006). Food Safety and Standards Act, 2006. Government of India. Retrieved from https://fssai.gov.in

Forret, M. L., & Dougherty, T. W. (2004). Networking behaviors and career outcomes: Differences for men and women? Journal of Organizational Behavior, 25(3), 419–437.

Gadgil, M., & Berkes, F. (1991). Traditional resource management systems. Resource Management and Optimization, 8(3-4), 127-141.

Gadgil, M., Hemam, N. S., & Reddy, B. M. (1998). People's Biodiversity Registers: A Methodology Manual. Centre for Ecological Sciences, Indian Institute of Science and National Biodiversity Strategy and Action Plan, Government of India.

Gassmann, O., Enkel, E., & Chesbrough, H. (2010). The future of open innovation. R&D Management, 40(3), 213–221. https://doi.org/10.1111/j.1467-9310.2010.00605.x

Geisler, E. (2000). The metrics of science and technology. Quorum Books.

Genetic Engineering Appraisal Committee. (2006). Regulations for Import of GM Products under Foreign Trade Policy. Ministry of Environment, Forest and Climate Change,

Goldscheider, R. (2001). Licensing and the 25% rule. Les Nouvelles – Journal of the Licensing Executives Society, 36(4), 123–126

Goldsmith, R. (2003). The Strategic Marketing of Science, Technology and Innovation. International Journal of Technology Management & Sustainable Development, 2(2), 103–118. https://doi.org/10.1386/ijtm.2.2.103/0

Government of India. (1872). The Indian Contract Act, 1872. Retrieved from https://legislative.gov.in/sites/default/files/A1872-9.pdf

Government of India. (1957). The Copyright Act, 1957 (as amended). Retrieved from https://copyright.gov.in

Government of India. (1957). The Copyright Act, 1957 (as amended). Retrieved from https://copyright.gov.in

Government of India. (1957). The Copyright Act, 1957 (as amended in 2012). Ministry of Law and Justice. Retrieved from https://copyright.gov.in/

Government of India. (1970). The Patents Act, 1970 (as amended). Retrieved from https://ipindia.gov.in

Government of India. (1970). The Patents Act, 1970 (as amended). Retrieved from https://ipindia.gov.in/writereaddata/Portal/ev/sections/ps.pdf

Government of India. (1970). The Patents Act, 1970 (as amended). Retrieved from https://ipindia.gov.in/writereaddata/Portal/IPOAct/1_31_1_patent-act-1970-11march2015.pdf

Government of India. (1970). The Patents Act, 1970 (as amended in 2005). Ministry of Law and Justice. Retrieved from https://legislative.gov.in/

Government of India. (1999). The Geographical Indications of Goods (Registration and Protection) Act, 1999. Ministry of Law and Justice. Retrieved from https://ipindia.gov.in/

Government of India. (1999). The Geographical Indications of Goods (Registration and Protection) Act, 1999. Retrieved from https://ipindia.gov.in/gir-act-rules.htm

Government of India. (1999). The Geographical Indications of Goods (Registration and Protection) Act, 1999. Retrieved from https://ipindia.gov.in/writereaddata/Portal/Images/pdf/GI_Act_1999.pdf

Government of India. (1999). The Geographical Indications of Goods (Registration and Protection) Act, 1999. Retrieved from https://legislative.gov.in/actsofparliamentfromtheyear/geographical-indications-goods- registration-and-protection-act-1999

Government of India. (1999). The Trade Marks Act, 1999 (as amended). Retrieved from https://ipindia.gov.in/trade-marks-act-rules.htm

Government of India. (1999). The Trade Marks Act, 1999 (as amended). Retrieved from https://ipindia.gov.in/writereaddata/Portal/ev/sections/tm.pdf

Government of India. (1999). The Trade Marks Act, 1999 (as amended in 2010). Ministry of Law and Justice. Retrieved from https://ipindia.gov.in/

Government of India. (2000). The Designs Act, 2000. Ministry of Law and Justice. Retrieved from https://ipindia.gov.in/

Government of India. (2000). The Designs Act, 2000. Retrieved from https://ipindia.gov.in/design-act-rules.htm

Government of India. (2000). The Designs Act, 2000. Retrieved from https://ipindia.gov.in/writereaddata/Portal/Images/pdf/Designs_Act.pdf

Government of India. (2000). The Semiconductor Integrated Circuits Layout-Design Act, 2000. Retrieved from https://ipindia.gov.in/semiconductor-integrated-circuits.htm

Government of India. (2000). The Semiconductor Integrated Circuits Layout-Design Act, 2000. Retrieved from https://legislative.gov.in/sites/default/files/A2000-37.pdf

Government of India. (2001). The Protection of Plant Varieties and Farmers' Rights (PPV&FR) Act, 2001. Retrieved from https://plantauthority.gov.in

Government of India. (2001). The Protection of Plant Varieties and Farmers' Rights Act, 2001. Ministry of Agriculture and Farmers Welfare. Retrieved from https://www.indiacode.nic.in

Government of India. (2001). The Protection of Plant Varieties and Farmers' Rights Act, 2001. Retrieved from https://plantauthority.gov.in

Government of India. (2002). The Biological Diversity Act, 2002. Ministry of Environment, Forest and Climate Change. Retrieved from https://nbaindia.org

Government of India. (2002). The Biological Diversity Act, 2002. Ministry of Environment, Forest and Climate Change. Retrieved from https://nbaindia.org/uploaded/pdf/Biodiversity_Act_2002.pdf

Government of India. (2002). The Biological Diversity Act, 2002. Ministry of Environment, Forest and Climate Change. Retrieved from https://www.moef.gov.in

Government of India. (2002). The Biological Diversity Act, 2002. Retrieved from https://nbaindia.org

Government of India. (2006). The Micro, Small and Medium Enterprises Development Act. Ministry of MSME.

Government of India. (2014). Make in India: A Major National Program. Ministry of Commerce and Industry. Retrieved from https://www.makeinindia.com/

Government of India. (2015). Digital India Programme. Ministry of Electronics and Information Technology. Retrieved from https://www.digitalindia.gov.in/

Government of India. (2015). National Policy on Skill Development and Entrepreneurship. Ministry of Skill Development and Entrepreneurship. Retrieved from https://www.msde.gov.in/

Government of India. (2016). Stand-Up India Scheme. Ministry of Finance. Retrieved from https://www.standupmitra.in/

Government of India. Retrieved from https://moef.gov.in

Granola Jr., E., et al. (1988). Technology assessment in mango production: Case study in the Philippines. Unpublished report.

Grimaldi, R., & Grandi, A. (2005). Business incubators and new venture creation: An assessment of incubating models. Technovation, 25(2), 111–121.

Grimaldi, R., & Grandi, A. (2005). Business incubators and new venture creation: An assessment of incubating models. Technovation, 25(2), 111–121.

Guffey, M. E., & Loewy, D. (2022). Business communication: Process and product (10th ed.). Cengage Learning.

Gupta, A. K. (2013). Tapping the Entrepreneurial Potential of Grassroots Innovation. Stanford Social Innovation Review. Retrieved from https://ssir.org/articles/entry/tapping_the_entrepreneurial_potential_of_grassroots_innovation

Gupta, A. K. (2016). Grassroots Innovation: Minds on the Margin Are Not Marginal Minds. New Delhi: Penguin Books.

Gupta, A. K. (2016). Grassroots innovation: Minds on the margin are not marginal minds. Penguin Books India.

Gupta, S. L., & Chaturvedi, M. (2020). Business communication: Skills, concepts and applications. Himalaya Publishing House.

Guston, D. H., & Sarewitz, D. (2002). Real-time technology assessment. Technology in Society, 24(1-2), 93–109. https://doi.org/10.1016/S0160-791X(01)00047-1

Hackett, S. M., & Dilts, D. M. (2004). A systematic review of business incubation research. The Journal of Technology Transfer, 29(1), 55–82. https://doi.org/10.1023/B:JOTT.0000011181.11952.0f

Hartmann, A., & Linn, J. F. (2008). Scaling up: A framework and lessons for development effectiveness from literature and practice. Wolfensohn Center for Development Working Paper No. 5. Brookings Institution.

Hisrich, R. D., Peters, M. P., & Shepherd, D. A. (2016). Entrepreneurship. McGraw-Hill Education.

Holcombe, S. (2012). Lessons from practice: Assessing scalability. Journal of Social Entrepreneurship, 3(1), 24–44.

Honey Bee Network. (n.d.). About Us. Retrieved from https://www.sristi.org/honeybee- network Ibarra, H., & Hunter, M. (2007). How leaders create and use networks. Harvard Business Review, 85(1), 40–47.

https://documents.worldbank.org/

https://doi.org/10.1002/job.253

https://doi.org/10.1016/j.bushor.2009.03.002

https://doi.org/10.1016/j.technovation.2016.02.005

https://doi.org/10.1016/S0048-7333(01)00148-X

https://doi.org/10.1016/S0048-7333(01)00152-6

https://doi.org/10.1016/S0166-4972(03)00076-2

https://doi.org/10.1016/S0166-4972(03)00076-2

https://doi.org/10.1016/S0883-9026(00)00055-0

https://doi.org/10.1016/S1574-0072(01)10007-1

https://doi.org/10.1080/09537320120098864

https://doi.org/10.1111/j.1467-6486.2008.00803.x

https://doi.org/10.1111/j.1467-6486.2008.00803.x

https://doi.org/10.1177/104225879001500207

https://doi.org/10.1287/mnsc.48.1.90.14271

https://doi.org/10.2139/ssrn.1393722

https://www.fao.org/3/ca6030en/ca6030en.pdf

https://www.wipo.int/edocs/pubdocs/en/wipo_pub_1043.pdf

https://www.wipo.int/lisbon/en/legal_texts/lisbon_agreement.html

https://www.wto.org/english/tratop_e/trips_e/pharmpatent_e.htm

ICAR. (2013). Handbook on Indigenous Technical Knowledge in Agriculture. Indian Council of Agricultural Research.

ICAR. (2018). Guidelines for Intellectual Property Management and Technology Transfer/ Commercialization. Indian Council of Agricultural Research (ICAR), New Delhi. Retrieved from https://icar.org.in/

ICRISAT. (n.d.). Agri-Business Incubator (ABI). International Crops Research Institute for the Semi-Arid Tropics (ICRISAT). Retrieved from https://www.icrisat.org/

Indian Council of Agricultural Research (ICAR). (1995). Manual on Farmer Participatory Technology Development and Dissemination Through Institute Village Linkage Programme (IVLP). Division of Agricultural Extension, ICAR, New Delhi.

Indian Council of Agricultural Research (ICAR). (2006). National Agricultural Innovation Project (NAIP). Retrieved from https://naip.icar.gov.in/

Indian Council of Agricultural Research (ICAR). (2020). Commercialization of Technologies: Success Stories and Guidelines. New Delhi: ICAR-NAARM. https://naarm.org.in

Indian Council of Agricultural Research (ICAR). (2022, July 8). a-IDEA of ICAR-NAARM signs MoA with 15 ICAR Institutes. Retrieved from https://naarm.org.in/

Indian Council of Agricultural Research. (1995). Institute-Village Linkage Programme (IVLP): Concept and guidelines. ICAR Publication.

Indian Council of Agricultural Research. (2006). ICAR Guidelines for Intellectual Property Management and Technology Transfer/Commercialization. New Delhi: ICAR. Retrieved from https://icar.org.in

Indian Council of Agricultural Research. (2006). ICAR Guidelines for Intellectual Property Management and Technology Transfer/Commercialization. New Delhi: ICAR. Retrieved from https://icar.org.in

Indian Council of Agricultural Research. (2019). ICAR Innovation and Startup Policy 2019. New Delhi: ICAR. Retrieved from https://icar.org.in/files/ICAR-Innovation-Startup-Policy-2019.pdf

Indian Institute of Management Ahmedabad. (n.d.). Centre for Innovation, Incubation and Entrepreneurship (CIIE.CO). Retrieved from https://www.ciie.co

Indian Institute of Technology Madras. (n.d.). Rural Technology and Business Incubator (RTBI). Retrieved from https://www.rtbi.in

Indian STEP and Business Incubator Association (ISBA). (n.d.). About Business Incubation in India. Retrieved from http://www.isba.in/

Innovations for Defence Excellence (iDEX), Ministry of Defence. (n.d.). iDEX Program. Retrieved from https://idex.gov.in

Intellectual Property India. (n.d.). Geographical Indications Registry. Retrieved from https://ipindia.gov.in/geographical-indications.htm

International Food Policy Research Institute. (2020). Digital Agriculture: Farmers in Africa are increasingly using technology. IFPRI. https://www.ifpri.org/blog/digital-agriculture-farmers-africa-are-increasingly-using-technology

International Service for the Acquisition of Agri-biotech Applications (ISAAA). (2020). Global Status of Commercialized Biotech/GM Crops: 2019. Retrieved from https://www.isaaa.org

International Union for the Protection of New Varieties of Plants (UPOV). (n.d.). About UPOV. Retrieved from https://www.upov.int/portal/index.html.en

International Valuation Standards Council (IVSC). (2021). International valuation standards (IVS) 2021. Retrieved from https://www.ivsc.org

IP Australia. (2018). Valuing Your Intellectual Property. Retrieved from https://www.ipaustralia.gov.au

IP Australia. (2018). Valuing your intellectual property. Retrieved from https://www.ipaustralia.gov.au

Jensen, R., & Thursby, M. (2001). Proofs and prototypes for sale: The licensing of university inventions. American Economic Review, 91(1), 240–259. https://doi.org/10.1257/aer.91.1.240

Kaiser, J. (2010). Co-creation and the protection of Intellectual Property. Journal of Intellectual Property Rights, 15(4), 278–284.

Kaiser, R. (2010). Patent strategy for researchers and research managers. Wiley-VCH.

Kane, G. C., Palmer, D., Phillips, A. N., Kiron, D., & Buckley, N. (2015). Strategy, Not Technology, Drives Digital Transformation. MIT Sloan Management Review and Deloitte University Press.

Kotler, P., & Keller, K. L. (2016). Marketing management (15th ed.). Pearson Education. Kotler, P., & Keller, K. L. (2016). Marketing management (15th ed.). Pearson Education.

KPMG. (2016). Valuation of intangible assets. Retrieved from https://home.kpmg/

Krattiger, A., Mahoney, R. T., Nelsen, L., Thomson, J. A., Bennett, A. B., Satyanarayana, K.,... & Koo, B. (2007). Intellectual Property Management in Health and Agricultural Innovation: A Handbook of Best Practices. MIHR, Oxford and PIPRA, Davis. Retrieved from https://www.iphandbook.org

Kumar, N., & Aggarwal, A. (2005). Intellectual Property Rights and Agricultural Technology: Interplay and Implications for India. Economic and Political Weekly, 40(3), 231–240. https://www.jstor.org/stable/4416095

Kumar, N., & Aggarwal, A. (2005). Liberalization, outward orientation and in-house R&D activity of multinational and local firms: A quantitative exploration for Indian manufacturing. Research Policy, 34(4), 441–460. https://doi.org/10.1016/j.respol.2005.01.005

Laboratories. R&D Management, 37(5), 503–513. https://doi.org/10.1111/j.1467-9310.2007.00423.x

Lee, K., & Win, H. N. (2004). Technology transfer between university research centers and industry in Singapore. Technovation, 24(5), 433–442. https://doi.org/10.1016/S0166-4972(02)00065-7

Linn, J. F. (2012). Scaling up in agriculture, rural development, and nutrition. 2020 Focus Brief. International Food Policy Research Institute (IFPRI).

Mangold, W. G., & Faulds, D. J. (2009). Social media: The new hybrid element of the promotion mix. Business Horizons, 52(4), 357–365.

Markham, S. K. (2002). Moving technologies from lab to market. Research-Technology Management, 45(6), 31–42. https://doi.org/10.1080/08956308.2002.11671542

Markman, G. D., Siegel, D. S., & Wright, M. (2008). Research and technology commercialization. Journal of Management Studies, 45(8), 1401–1423.

Markman, G. D., Siegel, D. S., & Wright, M. (2008). Research and technology commercialization. Journal of Management Studies, 45(8), 1401–1423.

McClean, M. (1988). A checklist approach to technology assessment. Agricultural Systems, 27(4), 295-310.

Mian, S. A. (1996). Assessing value-added contributions of university technology business incubators to tenant firms. Research Policy, 25(3), 325–335. https://doi.org/10.1016/0048-7333(95)00847-0

Mian, S. A., Lamine, W., & Fayolle, A. (2016). Technology Business Incubation: An overview of the state of knowledge. Technovation, 50-51, 1-12.

Microsoft Ventures. (n.d.). Microsoft for Startups Founders Hub. Retrieved from https://startups.microsoft.com

Ministry of Agriculture and Farmers Welfare. (2018). Innovation and Agri-Entrepreneurship Development Programme. Rashtriya Krishi Vikas Yojana (RKVY). Retrieved from https://rkvy.nic.in/

Ministry of AYUSH. (n.d.). Traditional Knowledge and AYUSH. Government of India. Retrieved April 7, 2025, from https://www.ayush.gov.in

Ministry of Commerce and Industry, Department for Promotion of Industry and Internal Trade (DPIIT). (n.d.). Office of the Controller General of Patents, Designs and Trade Marks (CGPDTM). Retrieved from https://ipindia.gov.in

Ministry of Commerce and Industry. (2006). Foreign Trade (Development and Regulation) Act, 1992 (as amended). Retrieved from https://commerce.gov.in

Ministry of Education. (2019). National Innovation and Startup Policy (NISP) 2019. All India Council for Technical Education. Retrieved from https://www.aicte-india.org/

Ministry of Electronics and Information Technology (MeitY). (n.d.). Technology Incubation and Development of Entrepreneurs (TIDE) 2.0 Scheme. Retrieved from https://www.meity.gov.in

Ministry of Environment, Forest and Climate Change (MoEFCC). (2004). Biological Diversity Rules, 2004. Retrieved from https://nbaindia.org/uploaded/pdf/BD_Rules_2004.pdf

Ministry of Environment, Forest and Climate Change (MoEFCC). (2014). Guidelines on Access to Biological Resources and Associated Knowledge and Benefit Sharing Regulations. National Biodiversity Authority. Retrieved from https://nbaindia.org

Ministry of Environment, Forest and Climate Change. (1986). The Environment (Protection) Act, 1986. Government of India. Retrieved from https://legislative.gov.in

Ministry of Environment, Forest and Climate Change. (2006). National Environment Policy (NEP). Government of India. Retrieved from https://moef.gov.in

Ministry of External Affairs, Government of India. (n.d.). BIMSTEC Overview. Retrieved from https://www.mea.gov.in/bimstec.htm

Ministry of Micro, Small and Medium Enterprises (MoMSME). (n.d.). ASPIRE Scheme for Promotion of Innovation, Rural Industries and Entrepreneurship. Retrieved from https://msme.gov.in

Ministry of Science and Technology, Government of India. (2022). National Innovation and Startup Policy (NISP). Retrieved from https://innovationcouncil.mic.gov.in/nisp/

Ministry of Science and Technology. (2009). Technology Business Incubators and NSTEDB Initiatives. Government of India. Retrieved from https://dst.gov.in/

Mishra, A., & Sahoo, B. K. (2020). Role of indigenous knowledge in climate change adaptation and mitigation: Perspectives from Indian agriculture. Indian Journal of Agricultural Sciences, 90(12), 2275–2282.

Misner, I. R., & Donovan, D. (2017). Networking like a pro: Turning contacts into connections (2nd ed.). Entrepreneur Press.

Misra, S. K. (1993). Agricultural development and environmental issues in India. In Brundtland Commission Report follow-up studies. New Delhi: Planning Commission.

MoEFCC. (2019). India's Sixth National Report to the Convention on Biological Diversity. Retrieved from https://www.cbd.int/doc/nr/nr-06/in-nr-06-en.pdf

NABARD. (2020). Producer Organization and SHG-Bank Linkage Program Reports. National Bank for Agriculture and Rural Development.

Narayanan, V. K. (2001). Managing technology and innovation for competitive advantage. Prentice Hall.

National Agricultural Innovation Project (NAIP). (2014). Innovations in Technology Commercialization. New Delhi: ICAR. Available at: https://naip.icar.gov.in

National Biodiversity Authority (NBA). (2004). Biological Diversity Act, 2002 and Biological Diversity Rules, 2004. Chennai: Ministry of Environment, Forests and Climate Change, Government of India. Retrieved from https://nbaindia.org

National Biodiversity Authority (NBA). (2020). Guidelines for Preparation of People's Biodiversity Registers (PBRs). Retrieved from https://nbaindia.org

National Biodiversity Authority (NBA). (n.d.). About the Biological Diversity Act, 2002. Retrieved April 7, 2025, from https://nbaindia.org/content/25/19/1/act.html

National Biodiversity Authority (NBA). (n.d.). Access and Benefit Sharing (ABS) Mechanism. Retrieved April 7, 2025, from https://nbaindia.org/content/25/21/1/abs.html

National Biodiversity Authority (NBA). (n.d.). Guidelines on Access to Biological Resources and Associated Knowledge and Benefits Sharing Regulations, 2014. Retrieved from https://nbaindia.org/uploaded/pdf/ABS_Guidelines.pdf

National Biodiversity Authority. (n.d.). ABS: Access and Benefit Sharing. Retrieved from https://nbaindia.org/content/27/20/1/abs.html

National Biodiversity Authority. (n.d.). Role of Biodiversity Management Committees (BMCs). Retrieved from https://nbaindia.org/content/21/61/1/biodiversity-management-committees- bmcs.html

National Business Incubation Association. (2020). What is business incubation? https://inbia.org/

National Innovation Foundation - India. (2017). Best Practices in Technology Transfer and IPR Management in Indian Research Institutions. Ahmedabad: NIF. Retrieved from https://nif.org.in

National Institutes of Health (NIH). (2020). Technology Transfer Process. Retrieved from https://www.ott.nih.gov

National Science & Technology Entrepreneurship Development Board (NSTEDB). (n.d.). Objectives and Programs. Retrieved from https://nstedb.com/

National Science and Technology Entrepreneurship Development Board (NSTEDB). (n.d.). About NSTEDB. Department of Science and Technology, Government of India. Retrieved from https://dst.gov.in/

National Science and Technology Entrepreneurship Development Board (NSTEDB). (n.d.). Science and Technology Entrepreneurs Parks (STEPs) and Technology Business Incubators (TBIs). Retrieved from https://nstedb.com

National Science Foundation. (2020). Technology transfer and commercialization landscape. Retrieved from https://www.nsf.gov

Nestel, B. (1989). Technology assessment methodology for agricultural development. Food Policy, 14(2), 135–142.

OECD. (2004). Patents and innovation: Trends and policy challenges. Organisation for Economic Co-operation and Development. Retrieved from https://www.oecd.org

OECD. (2006). Guidelines for the Licensing of Genetic Inventions. Organisation for Economic

OECD. (2008). Intellectual Assets and Value Creation: Implications for Corporate Reporting. Organisation for Economic Co-operation and Development. Retrieved from https://www.oecd.org

OECD. (2011). Fostering Innovation to Address Social Challenges. OECD Publishing. https://doi.org/10.1787/9789264126269-en

OECD. (2013). Commercialising Public Research: New Trends and Strategies. Paris: OECD Publishing. https://doi.org/10.1787/9789264193321-en

Organisation for Economic Co-operation and Development (OECD). (2003). Turning Science into Business: Patenting and Licensing at Public Research Organisations. Paris: OECD Publishing. https://doi.org/10.1787/9789264100244-en

Organisation for Economic Co-operation and Development (OECD). (2005). Oslo manual: Guidelines for collecting and interpreting innovation data (3rd ed.). OECD Publishing. https://doi.org/10.1787/9789264013100-en

Organisation for Economic Co-operation and Development (OECD). (2011). Guidelines on the transfer pricing aspects of intangibles. OECD Publishing. Retrieved from https://www.oecd.org/

Parr, R. L., & Smith, G. V. (2000). Valuation of intellectual property and intangible assets (3rd ed.). Wiley.

Phillips, R. G. (2002). Technology business incubators: How effective as technology transfer mechanisms? Technology in Society, 24(3), 299–316. https://doi.org/10.1016/S0160-791X(02)00010-6

Planning Commission. (2002). Tenth Five Year Plan (2002-2007): Volume II. Government of India. Retrieved from http://planningcommission.nic.in/

Porter, M. E. (1985). Competitive Advantage: Creating and Sustaining Superior Performance. Free Press.

Porter, M. E. (1998). Clusters and the New Economics of Competition. Harvard Business Review, 76(6), 77–90.

Powell, M. C., & Colin, M. (2008). Meaningful citizen engagement in science and technology: What would it really take? Science Communication, 30(1), 126–136. https://doi.org/10.1177/1075547008316222

PPV&FR Authority. (2020). Annual Report 2019-20. Ministry of Agriculture and Farmers Welfare. Retrieved from http://www.plantauthority.gov.in/

Pratt, S. P., Niculita, A. V., & Grabowski, R. J. (2008). Valuing a business: The analysis and appraisal of closely held companies (5th ed.). McGraw-Hill.

Press Information Bureau. (2023). ICAR signs MOU with Dhanuka Agritech Ltd to benefit smallholder farmers. Government of India. Retrieved from https://pib.gov.in/

Protection of Plant Varieties and Farmers' Rights Authority (PPVFRA). (2001). The Protection of Plant Varieties and Farmers' Rights Act, 2001. Government of India.

Rao, M. R., & Raju, K. V. (2014). Agribusiness and Innovation Systems in India. International Food Policy Research Institute (IFPRI).

Reddy, G. P. O., & Rao, P. P. (2017). Indigenous knowledge and natural resource management: A study among the tribes in Andhra Pradesh. Indian Journal of Traditional Knowledge, 16(1), 122–128.

Reilly, R. F., & Schweihs, R. P. (2004). Guide to Intangible Asset Valuation. American Institute of Certified Public Accountants.

Reilly, R. F., & Schweihs, R. P. (2004). Valuing intangible assets. McGraw-Hill.

Renn, O., Webler, T., Rakel, H., Dienel, P., & Johnson, B. (1993). Public participation in decision making: A three-step procedure. Policy Sciences, 26, 189–214. https://doi.org/10.1007/BF00999716

Rice, M. P. (2002). Co-production of business assistance in business incubators: An exploratory study. Journal of Business Venturing, 17(2), 163–187.

Ries, E. (2011). The Lean Startup: How Today's Entrepreneurs Use Continuous Innovation to Create Radically Successful Businesses. Crown Publishing.

Rogers, E. M. (2003). Diffusion of innovations (5th ed.). Free Press.

Rogers, E. M. (2003). Diffusion of innovations (5th ed.). Free Press. Rogers, E. M. (2003). Diffusion of innovations (5th ed.). Free Press.

Rohrbeck, R. (2007). Technology scouting: A case study on the Deutsche Telekom

Ross, J. W., Sebastian, I. M., & Beath, C. M. (2019). How to develop a great digital strategy. MIT Sloan Management Review, 60(2), 7–9.

Rules for the Manufacture, Use, Import, Export and Storage of Hazardous Microorganisms/Genetically Engineered Organisms or Cells. (1989). Under the Environment (Protection) Act, 1986. Ministry of Environment and Forests, Government of India. Retrieved from https://envfor.nic.in

Saha, S. (2011). Geographical Indications: A Tool for Economic and Cultural Development in India. Indian Journal of Economics and Development, 7(2), 23–29.

Sampat, B. N. (2009). The Bayh-Dole model in developing countries: Reflections on the Indian bill on publicly funded intellectual property. Policy Brief, UNCTAD–ICTSD Project on IPRs and Sustainable Development. Geneva: UNCTAD.

Sampat, B. N. (2009). The Bayh-Dole model in developing countries: Reflections on the Indian bill on publicly funded intellectual property. UNU-MERIT Working Paper Series, 2009-015.

Sanyang, S. E., & Huang, W. C. (2008). Entrepreneurship and economic development: The role of entrepreneurship education in developing countries. International Journal of Education and Development using ICT, 4(3). https://www.learntechlib.org/p/42223/

Sarasvathy, S. D. (2001). Causation and effectuation: Toward a theoretical shift from economic inevitability to entrepreneurial contingency. Academy of Management Review, 26(2), 243–263.

Schawbel, D. (2013). Promote yourself: The new rules for career success. St. Martin's Press.

Schilling, M. A. (2020). Strategic Management of Technological Innovation (6th ed.). McGraw-Hill Education.

Schilling, M. A. (2020). Strategic management of technological innovation (6th ed.). McGraw-Hill Education.

Schumpeter, J. A. (1934). The Theory of Economic Development. Harvard University Press.

Secretariat of the Convention on Biological Diversity. (2011). Nagoya Protocol on Access to Genetic Resources and the Fair and Equitable Sharing of Benefits Arising from Their Utilization. United Nations. Retrieved from https://www.cbd.int/abs/

Secretariat of the Convention on Biological Diversity. (n.d.). About the Convention. Retrieved from https://www.cbd.int/

Seedfund. (n.d.). India's early-stage venture capital firm. Retrieved from http://www.seedfund.in/

Seetharaman, A. (2009). Communication skills for effective business management. Journal of Contemporary Management Research, 3(1), 1–13.

SELCO India. (n.d.). Case studies of rural impact through renewable energy. Retrieved from https://www.selco-india.com/

Selin, C., & Sadowski, J. (2016). Citizens' perspectives in technology assessment: A critical review of the literature. Science and Public Policy, 43(3), 345–358. https://doi.org/10.1093/scipol/scv076

Shane, S. (2004). Academic Entrepreneurship: University Spinoffs and Wealth Creation. Cheltenham, UK: Edward Elgar Publishing.

Shiva, V. (2000). Stolen Harvest: The Hijacking of the Global Food Supply. South End Press.

Siegel, D. S., Waldman, D., & Link, A. N. (2003). Assessing the impact of organizational practices on the relative productivity of university technology transfer offices: An exploratory study. Research Policy, 32(1), 27–48. https://doi.org/10.1016/S0048-7333(01)00196-2

Siegel, D. S., Waldman, D., & Link, A. N. (2003). Assessing the impact of organizational practices on the relative productivity of university technology transfer offices: An exploratory study. Research Policy, 32(1), 27–48. https://doi.org/10.1016/S0048-7333(01)00196-2

Siegel, D. S., Waldman, D. A., & Link, A. N. (2003). Assessing the impact of organizational practices on the productivity of university technology transfer offices: An exploratory study. Research Policy, 32(1), 27–48. https://doi.org/10.1016/S0048-7333(01)00196-2

Siegel, D. S., Waldman, D. A., & Link, A. N. (2003). Assessing the impact of organizational practices on the relative productivity of university technology transfer offices: An exploratory study. Research Policy, 32(1), 27–48. https://doi.org/10.1016/S0048-

7333(01)00196-2

Sillitoe, P. (2000). Indigenous knowledge development in the South: Contesting knowledge and practice. Zed Books.

Singh, B. K., & Meena, M. S. (2012). Participatory technology development and assessment for sustainable agriculture. Indian Research Journal of Extension Education, 12(3), 1–6.

Singh, K. M. (1996). Appropriateness of technology and its impact on technology adoption. Indian Journal of Agricultural Economics, 51(3), 398-403.

Smith, G. V., & Parr, R. L. (2005). Intellectual Property: Valuation, Exploitation, and Infringement Damages (3rd ed.). John Wiley & Sons.

Smith, G. V., & Parr, R. L. (2005). Intellectual property: Valuation, exploitation, and infringement damages (3rd ed.). Wiley.

Smith, G. V., & Parr, R. L. (2005). Intellectual property: Valuation, exploitation, and infringement damages (3rd ed.). Wiley.

Smith, P. R. & Zook, Z. (2016). Marketing communications: Offline and online integration, engagement and analytics (6th ed.). Kogan Page Publishers.

Society for Innovation and Entrepreneurship, IIT Bombay. (n.d.). SINE IIT Bombay. Retrieved from https://www.sineiitb.org

Society for Research and Initiatives for Sustainable Technologies and Institutions (SRISTI). (n.d.). About SRISTI. Retrieved from https://www.sristi.org

Society for Technology Management (STEM). (n.d.). Promoting Technology Transfer in Public Research Systems. Retrieved from http://www.stemglobal.org/

Stand-Up India. (n.d.). Stand-Up India Scheme for SC/ST and Women Entrepreneurs. Retrieved from https://www.standupmitra.in

Startup India. (n.d.). Government support for startups. Retrieved from https://www.startupindia.gov.in/

Startup India. (n.d.). Startup India Seed Fund Scheme. Retrieved from https://www.startupindia.gov.in

Startup Village. (n.d.). India's First PPP Model Startup Incubator. Retrieved from https://www.startupvillage.in

Stilgoe, J., Owen, R., & Macnaghten, P. (2013). Developing a framework for responsible innovation. Research Policy, 42(9), 1568–1580. https://doi.org/10.1016/j.respol.2013.05.008

Strasser, H. (1972). Technology assessment: A critical appraisal. Science, Technology & Human Values, 9(1), 1-12.

Sullivan, P. H. (2000). Value-Driven Intellectual Capital: How to Convert Intangible Corporate Assets into Market Value. John Wiley & Sons.

Sullivan, P. H. (2000). Value-driven intellectual capital: How to convert intangible corporate assets into market value. Wiley.

Sunding, D., & Zilberman, D. (2001). The agricultural innovation process: Research and technology adoption in a changing agricultural sector. In B. Gardner & G. Rausser (Eds.), Handbook of Agricultural Economics (Vol. 1, pp. 207–261). Elsevier.

Swaminathan, M. S. (2002). Agrobiodiversity and Farmers' Rights. In S. A. Noronha & N. P. Singh (Eds.), Protecting the Rights of Farmers in the Emerging Context of Intellectual Property Regimes in India (pp. 25–45). M.S. Swaminathan Research Foundation.

Swaminathan, M. S. (2007). Agricultural Research and Intellectual Property Rights. Indian Journal of Agricultural Sciences, 77(9), 519–525.

Syngenta India. (2010). Sunflower hybrid SB-207 performance report and farmer testimonials. Retrieved from https://www.syngenta.in

Teece, D. J. (2010). Business models, business strategy and innovation. Long Range Planning, 43(2–3), 172–194. https://doi.org/10.1016/j.lrp.2009.07.003

Thursby, J. G., & Kemp, S. (2002). Growth and productive efficiency of university intellectual property licensing. Research Policy, 31(1), 109–124. https://doi.org/10.1016/S0048-7333(01)00108-6

Thursby, J. G., & Thursby, M. C. (2002). Who is selling the ivory tower? Sources of growth in university licensing. Management Science, 48(1), 90–104.

Tidd, J., & Bessant, J. (2018). Managing innovation: Integrating technological, market and organizational change (6th ed.). Wiley.

Tripp, R., Louwaars, N., & Eaton, D. (2007). Plant Variety Protection in Developing Countries: A Report from the Field. Food Policy, 32(3), 354–371. https://doi.org/10.1016/j.foodpol.2006.09.002

Tyler, K., & Davies, S. (2022). Etiquette in business communication. In S. Davies (Ed.), Professional communication skills for business (pp. 55–78). Routledge.

U.S. Department of Commerce. (2013). Federal laboratory technology transfer summary report to the president and the congress: Fiscal years 2011–2012. National Institute of Standards and Technology. Retrieved from https://www.nist.gov

United Nations Convention to Combat Desertification (UNCCD). (1994). Article 16 – Access to Technology. Retrieved from https://www.unccd.int/

United Nations Development Programme (UNDP) India. (2012). Strengthening the implementation of the Biological Diversity Act and Rules with focus on its Access and Benefit Sharing Provisions. Retrieved from https://www.in.undp.org

United Nations Educational, Scientific and Cultural Organization (UNESCO). (2009). Learning and knowing in indigenous societies today. UNESCO.

United Nations Environment Programme (UNEP). (n.d.). Convention on Biological Diversity (CBD). Retrieved from https://www.unep.org/

United States Trade Representative (USTR). (2016). 2016 Special 301 Report. Retrieved from https://ustr.gov/sites/default/files/USTR-2016-Special-301-Report.pdf

UPOV. (1991). International Convention for the Protection of New Varieties of Plants. Retrieved from https://www.upov.int/upovlex/en/conventions/1991/content.html

Van de Ven, A. H. (1986). Central problems in the management of innovation. Management Science, 32(5), 590–607. https://doi.org/10.1287/mnsc.32.5.590

Wadhwani Foundation. (n.d.). Entrepreneurship and incubator programs in India. Retrieved from https://www.wfglobal.org/

Wagner, K., & Watch, G. (2006). The Technology Radar – Edition II. Deutsche Telekom Laboratories.

Warren, D. M., Slikkerveer, L. J., & Brokensha, D. (Eds.). (1995). The cultural dimension of development: Indigenous knowledge systems. Intermediate Technology Publications.

Watal, J. (2001). Intellectual property rights in the WTO and developing countries. Oxford University Press.

Weinberger, D. (2011). Too big to know: Rethinking knowledge now that the facts aren't the facts, experts are everywhere, and the smartest person in the room is the room. Basic Books.

Westerman, G., Bonnet, D., & McAfee, A. (2014). Leading Digital: Turning Technology into Business Transformation. Harvard Business Review Press.

Wilsdon, J., & Willis, R. (2004). See-through Science: Why public engagement needs to move upstream. Demos.

WIPO (World Intellectual Property Organization). (n.d.). Intellectual Property and Traditional Knowledge. Retrieved from https://www.wipo.int/tk/en/

WIPO. (1883). Paris Convention for the Protection of Industrial Property. Retrieved from https://www.wipo.int/treaties/en/ip/paris/

WIPO. (1886). Berne Convention for the Protection of Literary and Artistic Works. Retrieved from https://www.wipo.int/treaties/en/ip/berne/

WIPO. (2004). Exchanging value: Negotiating technology licensing agreements. World Intellectual Property Organization. Retrieved from https://www.wipo.int/publications/en/details.jsp?id=772

WIPO. (2011). IP Management at Universities and Research Institutions: A Guide to Developing Institutional IP Policy. Geneva: World Intellectual Property Organization. Retrieved from https://www.wipo.int/publications/en/details.jsp?id=336

WIPO. (2015). Guide to intellectual property for universities and R&D institutions. World Intellectual Property Organization. Retrieved from https://www.wipo.int

WIPO. (2015). Intellectual Property Handbook: Policy, Law and Use (2nd ed.). World Intellectual Property Organization. Retrieved from https://www.wipo.int/edocs/pubdocs/en/intproperty/489/wipo_pub_489.pdf

WIPO. (2020). Guide to Intellectual Property Valuation. World Intellectual Property Organization. Retrieved from https://www.wipo.int

WIPO. (2020). Managing intellectual property for universities. World Intellectual Property Organization. Retrieved from https://www.wipo.int

World Bank. (2006). Project Appraisal Document on a Proposed Loan to the Republic of India for a National Agricultural Innovation Project. Report No. 34427-IN. Retrieved from

World Bank. (2014). Promoting entrepreneurship and innovative SMEs in a global economy: Towards a more responsible and inclusive globalization. https://www.worldbank.org

World Bank. (2017). ICT in Agriculture: Connecting Smallholders to Knowledge, Networks, and Institutions (Updated ed.). The World Bank. https://doi.org/10.1596/978-1-4648-1002-2

World Health Organization. (1978). Declaration of Alma-Ata. International Conference on Primary Health Care. Retrieved from https://www.who.int/publications/almaata_declaration_en.pdf

World Health Organization. (2005). Modern food biotechnology, human health and development: An evidence-based study. Retrieved from https://www.who.int

World Intellectual Property Organization (WIPO). (2015). A Guide to Technology Transfer: How to Make the Most of Your Intellectual Property. Geneva: WIPO. https://www.wipo.int/edocs/pubdocs/en/wipo_pub_1039.pdf

World Intellectual Property Organization (WIPO). (2015). IP valuation: What methods are used to value intellectual property? Retrieved from https://www.wipo.int

World Intellectual Property Organization (WIPO). (2017). Documenting traditional knowledge – A toolkit. Geneva: WIPO. https://www.wipo.int/edocs/pubdocs/en/wipo_pub_1049.pdf

World Intellectual Property Organization (WIPO). (2020). Managing Intellectual Property in the Innovation Process: A Guide for Inventors and Researchers. Geneva: WIPO. Retrieved from https://www.wipo.int/publications/en/details.jsp?id=4536

World Intellectual Property Organization (WIPO). (n.d.). Geographical Indications. Retrieved from https://www.wipo.int/geo_indications/en/

World Intellectual Property Organization (WIPO). (n.d.). Intellectual Property and Commercialization. Retrieved from https://www.wipo.int/

World Intellectual Property Organization (WIPO). (n.d.). Lisbon Agreement for the Protection of Appellations of Origin and their International Registration. Retrieved from

World Intellectual Property Organization (WIPO). (n.d.). What is Intellectual Property?. Retrieved from https://www.wipo.int/about-ip/en/

World Intellectual Property Organization (WIPO). (n.d.). What is Intellectual Property? Retrieved from https://www.wipo.int/about-ip/en/

World Intellectual Property Organization. (2017). Protecting Traditional Knowledge: A Practical Guide for Policymakers. Geneva: WIPO. Retrieved from

World Intellectual Property Organization. (n.d.). Traditional Knowledge. Retrieved April 7, 2025, from https://www.wipo.int/tk/en/

World Trade Organization (WTO). (1994). Agreement on Trade-Related Aspects of Intellectual Property Rights (TRIPS). Retrieved from https://www.wto.org/english/tratop_e/trips_e/trips_e.htm

World Trade Organization (WTO). (1995). Agreement on Trade-Related Aspects of Intellectual Property Rights (TRIPS). Retrieved from https://www.wto.org/english/tratop_e/trips_e/trips_e.htm

World Trade Organization (WTO). (2001). Declaration on the TRIPS Agreement and Public Health. Retrieved from https://www.wto.org/english/thewto_e/minist_e/min01_e/mindecl_trips_e.htm

World Trade Organization (WTO). (n.d.). Trade-Related Aspects of Intellectual Property Rights (TRIPS). Retrieved from https://www.wto.org/english/tratop_e/trips_e/trips_e.htm

World Trade Organization. (1994). Agreement on Trade-Related Aspects of Intellectual Property Rights (TRIPS). Retrieved from https://www.wto.org/english/docs_e/legal_e/27- trips.pdf

World Trade Organization. (2001). Doha Ministerial Declaration. Retrieved from https://www.wto.org/english/thewto_e/minist_e/min01_e/mindecl_e.htm

World Wide Fund for Nature – India (WWF India). (n.d.). Biodiversity Conservation Prioritisation Project. Retrieved from https://www.wwfindia.org

WTO. (n.d.). TRIPS and public health. Retrieved from

Zhang, Y., Wang, G., & Duan, Y. (2016). Agricultural information dissemination using ICTs: A review and analysis. Information Processing in Agriculture, 3(1), 17–29. https://doi.org/10.1016/j.inpa.2015.11.002

Index